OUR
UNIVERSE

OUR UNIVERSE

A Scientific
and Religious View of Creation

FAIZ M. KHAN

iUniverse, Inc.
New York Lincoln Shanghai

OUR UNIVERSE
A Scientific and Religious View of Creation

iUniverse books may be ordered through booksellers or by contacting:

iUniverse
2021 Pine Lake Road, Suite 100
Lincoln, NE 68512
www.iuniverse.com
1-800-Authors (1-800-288-4677)

Because of the dynamic nature of the Internet, any Web addresses or links contained in this book may have changed since publication and may no longer be valid.

The views expressed in this work are solely those of the author and do not necessarily reflect the views of the publisher, and the publisher hereby disclaims any responsibility for them.

ISBN: 978-0-595-43006-2 (pbk)
ISBN: 978-0-595-87347-0 (ebk)

Printed in the United States of America

CONTENTS

Part 4   *Intellectual Challenges*

Part 5   *At The Crossroads Of Faith And Reason*

PREFACE

I dedicate this book to my daughters who did not inherit the religion of their father or their mother. Instead, I thought they should inherit this book.

The mother of my children is a Christian, and I am a Muslim, both by birth. When we married, we did not think religion would be an issue in our marriage. We had an undeclared understanding that our children would learn about religion when they grew up or by osmosis. No Sunday schools or madrassas were contemplated as part of their early educations. Now that they are grown up, educated, and engaged in their professional careers, I think religion did not play a significant role in shaping their lives. They follow moral principles of the society and culture like other decent people, nonetheless, but without declaring any religious label or affiliation. Because they were not instructed in their parents' religions when they were young, I thought I could write something now for them about my beliefs and life in general.

But I soon realized that my religious ideas were very superficial and not fit for communicating to my children. I did not feel comfortable with the thought that I believed in God because I was brought up with that belief. I needed a better rationale than that. At the same time that I was questioning the rationality of my inherited faith, I was wrestling with the notion of God as the creator of the universe. These questions prompted me to undertake a serious search for answers. The search had to be knowledge-based, so I started the quest with an in-depth study of cosmology and world religions. At the end of the quest, I felt elated by the knowledge I had gained. It was not long before I entertained the idea of writing this book so that I could share my experience with others. And here it is.

I have organized the book into five parts. Part 1 describes my early experiences with religion and my self-awareness as an individual human being. Part 2 is about scientific reasoning. For the benefit of the readers, I devote a couple of chapters in this part to cover the basic principles of scientific thought and the fundamental laws of physics underlying the cosmic theories. This discussion is elementary, approximately at the level of high school or early college science. The next four chapters provide current scientific views on how our universe came into existence, how it evolved, and how life originated and thrived on Earth. This part of the book is intended to give the reader a rational understanding of how the universe could have been created spontaneously, without cause.

Once the reader has grasped the scientific basis of creation, it is time to explore what other ideas exist that can compete with the scientific theories. Of course, religion has been preoccupied with creation philosophy ever since the beginning of the human race. So I carry my quest, along with the reader, into examining these philosophies and beliefs that are prevalent among nine major religions of the world—Hinduism, Buddhism, Sikhism, Taoism, Confucianism, Shintoism, Judaism, Christianity, and Islam. These are presented in part 3—the quest for God through faith.

In writing part 3, I focused primarily on the concept of God. By narrowing it down to the object of my quest, I wanted to concentrate on the reality of God or the concept of a supreme being as explained in the various scriptures.

From the very beginning, I was determined to be unbiased. Accordingly, I tossed all my prejudices out the window. Although I realize that one cannot erase years of brainwashing that we are all subjected to throughout our lives, I felt that it was critical to the honesty of my quest that I expunge as much as possible any preconceived notions of God and the related beliefs from my mind. I looked at each religion that I studied with the same regard as I had for my own religion or any other philosophy that I subscribe to.

The most difficult problem with all religions is their embrace of supernaturalism without question. God is supernatural. He is invisible. (In this book God is arbitrarily referred to as masculine. It is not intended to contradict the belief that God is spirit and has no gender.) You cannot experiment with him. His existence cannot be proved or disproved. We learn about him from the prophets and sages, but there is no direct confirmation of his existence. Scriptures present the most powerful testimony, but it is a testimony by men in the ancient past. We cannot cross-examine the authors. Scriptural literature is beautiful. It can grab your heart and convince you of the truth of God without proof. But rational thinking would tell you that it is all emotion, belief, and trust in the unknown. Critics would call it superstition.

Part 4 discusses intellectual challenges to religion. Agnosticism, atheism, humanism, secular humanism, and religious humanism present alternative philosophies based on scientific reasoning. Fairly or unfairly, they are often criticized as being godless and, therefore, without morality or purpose in life. These questions are openly debated in this part of the book, pitting secular views of life against the religious views in which God is the centerpiece.

At this point in the book, it is hoped that the reader will be well acquainted with the scientific and religious ideas regarding God, creation, evolution, and competing philosophies of life. Equipped with this background knowledge, the reader is standing at the crossroads of faith and reason. In part 5 it is time to engage in the debate over the existence of God. Because the background material has already been covered, all issues regarding creation, evolution, morality, and God are now laid out on the table. After a hearty discourse, we get to chapter 15, which marks the end of my quest, followed by a declaration of self-enlightenment. In the last chapter, I expound my personal philosophy of life and the principles of enlightenment that I have gleaned from this quest.

The quest for God and his role in the creation of our universe has been a great learning experience for me. It has given me an in-depth understanding of how our universe was most probably created and a true appreciation of the differences between faith and science. Most importantly, this study has given me confidence that, as a human being, I am empowered to lead my life as I see fit without fear of death or disposition after death. With knowledge and rationality, I can determine my own destiny, beliefs, and purpose in life.

I acknowledge Michael Fiedler, Publishing Services Associate, and the editorial staff of iUniverse for their guidance and useful suggestions.

Finally, I greatly appreciate my wife, Kathy. She was a frequent sounding board during my formulation of dialogues between faith and science. I also acknowledge my friends who were gracious enough to listen to some of my unconventional ideas and not mind it even when light conversations turned into serious debates or disagreements.

Faiz M. Khan

PART 1

EARLY IMPRESSIONS

CHAPTER 1
FIRST AWARENESS

I have no recollection of my birth or of anything that happened during my infancy until maybe a year or so after birth. It seems all of us are born with a clear memory bank. The potential is there, but the brain at this stage cannot process information to the extent that it can be stored in memory for later recollection. Although a baby can react to physical stimuli even in the womb, it cannot sustain the memory of its experiences until the brain has developed and matured sufficiently. Even then, the process is so gradual and continuous that most people have a hard time remembering the first moment they became aware of themselves or their environment.

Although there appears to be no eureka moment of first awareness, one could recollect an event in early childhood that might qualify to be the first moment of awareness in one's life. Because such an event is closely tied to memory, its recollection is more a function of one's long-term memory strength than the exact timeliness of its occurrence. For example, in my case, I can recollect my first awareness as a moment when I peeked through my baby sack (something like a Pak-A-Poose) carried on the back of someone (perhaps my mother). I saw a huge red ball of light in the far distance (maybe a setting or rising sun on the horizon). I could also discern the back of a person at a short distance (could be my father). This earliest snapshot of my self-awareness and of the surroundings has vividly stayed in my memory since my very early childhood. I do not know how old I was at the time of my first awareness, but certainly a baby peeking out of a Pak-A-Poose could not have been much older than a year.

I have used the term "awareness" here to imply recognition, perception, or the processing of the visual stimulus into an image and the storing of it in one's memory. At the time of first awareness, of course, the child has only a rudimentary capability of pattern recognition or interpretation of visual stimuli. It is also possible that at such an early stage of brain development, the information is stored with minimal processing at the start but can further mature into a more recognizable pattern as the child grows. In either case, it is fascinating to be able to remember the first time one is able to feel as an individual and make some sense out of the external and internal stimuli.

As human beings develop their awareness of the self and the environment, their other mental faculties grow, enabling them to think, reflect, and form ideas and opinions. Family, friends, community, and culture have an immense effect on mental growth—so much so that the process almost amounts to a gradual programming of the mind to form perceptions, beliefs, and mental attitudes later in life. However, individual differences exist in the genetic code that are also responsible for providing the variety we observe in human behaviors and responses, even if subjected to the same cultural environments.

The processes of mental development and cultural programming continue throughout one's life. As we reflect back on our lives, we find a dynamic process at work in which our ideas and opinions are formed, modified, and molded by the environment. The early childhood experiences are interesting because this is when the mental development and conditioning is occurring at its fastest pace. Young minds are known to be more impressionable than their older counterparts. The events and experiences at this stage of life get amplified and have a profound influence on the child's intelligence, character, social behavior, and intellectual pursuits. Seeds of human behavior are sown very early in life.

Quest for knowledge and understanding of the environment also starts very early. The child is curious about everything new and untiringly seeks answers to the enigmas presented in life at every step. The answers don't have to be rational or correct as long as they satisfy the child's curiosity. Myths are just as easily accepted as any scientific truth.

I can recall many irrational answers that I accepted as truths in my early childhood. For example, I was told that the earth was supported on the horns of a bull. Once in a while, the bull shifted the earth from one horn to the other and that caused the earthquakes. The answer to what caused rain was that the angels poured water into the clouds causing rain and thunder. Lightning was the fiery whip of the angel that drove the clouds to various destinations. These answers and explanations were not necessarily coming from my parents or my close family. In most cases, the myths and beliefs were perpetrated by my peers and elders who enjoyed telling stories and myths to little children.

Mythical and irrational beliefs began to crumble when I learned in school that the earth was a planet that circled the sun by mutual gravity, that earthquakes were caused by movements of Earth's plates, that rain was the result of condensation of water vapor, and that thunder was the noise created by lightning—an electric discharge between clouds or a cloud and an object on Earth.

Human beings are endowed with a powerful brain that is capable of thinking, reflecting, reasoning, and unraveling many mysteries of the universe. But it has to accumulate basic knowledge first before it can question, theorize, or confirm the truths it is seeking. The database for in-depth knowledge mostly resides in our schools, libraries, published literature, computers, and the minds of the learned scholars. We are fortunate that at this stage of human development, we have enormous resources at our disposal to help explore the universe and seek the truth about its origin, existence, and even its creator.

In my early schooling, I was exposed to religion, science, and arts. I took religion for granted because I was not expected to question it or experiment with it. It was ancient history, and I was supposed to believe in its every fable, injunction, or commandment. Not only was I to believe it but also practice it in daily life and affirm it socially through organized rituals and ceremonies. I inherited religion from my parents. I found no hardship in accepting it or believing it from the bottom of my heart. Although there were some occasional differences of opinion, mostly sectarian in nature, they were usually resolved in my mind, mostly in favor of my original beliefs. Religion was part of my life and came as natural to me as eating, drinking, or breathing. It was not until very late in life that I began to think seriously about religion—in fact, all religions—and to search for the truth about God's existence.

Next to religion, I enjoyed the arts because they are fun, thought provoking, and a tonic for our emotional well-being. Art bears a close relationship with religions but without their constraints or taboos. Freedom of thought and imagination is the hallmark of art. Its beauty is more in its form than substance, and its appeal is more to emotions than intellect. In fact, art helped me to appreciate religion without proving its authenticity. It nurtured my creative abilities to pursue science, and most importantly, it helped me to understand and appreciate my emotions.

In my early schooling, I fell in love with math and physics. I do not know why I was so enamored by mathematical theorems and their proofs. Maybe it was because mathematics represented an absolute truth to me—a truth unadulterated by conjecture, hearsay, or opinion. Its precision and exactness is unmatched by any other class of human knowledge. Its absoluteness almost approaches God's.

In spite of my extensive schooling in math and science, I have not been able to understand a few quandaries of mathematics. What is absolute zero, and what is infinity? They are the two extreme conditions that the human mind has not been able to comprehend. And yet we do not give up trying. Most people understand zero as nothingness, a complete absence of anything or a void. But can we attain a state of absolute nothingness in the real world? I think understanding zero and infinity might help us understand God. We will discuss this concept later in the book.

While math is the mother of all sciences, physics is its first offspring. If we combine the two, we can explain all other sciences: chemistry, biology, engineering, computers, and so forth. In fact, we can explain the whole universe with math and physics. But can we answer the ultimate question: who created the universe or how was the universe created?

To me, physics was the most exciting subject I studied in school. I was intrigued by what we call the laws of physics. I could appreciate Newton's laws of motion, law of gravity, and laws of conservation of energy and momentum, as well as theories on the nature of light, radiation, and fundamental particles of matter. But there were always these nagging questions lurking in the back of my mind that remained unanswered: Who created these laws? How did these laws come about? Is physical reality a consequence of these laws, or is our earthly experience nothing but a physical property of our brain?

After an in-depth study of the human brain and how it functions, I cannot escape the conclusion that all brain functions can be explained by physics and math. We may not understand many of its intricacies, but there is no mystery about what underlies its function and power. Constitutionally, the brain is made up of physical stuff. Like all physical systems, it obeys the laws of physics and math as it tries to understand them. But, outside these laws, is there anything extraneous that the brain is infused with? Is there a "spirit" or "soul" that turns the brain into what we call a mind? We will explore these questions further in this book as we continue our search for the truth.

Introduction to God

I do not recollect the precise moment when I first heard the word "God," but I do remember times in my early childhood when God's name was brought up or when explanations were given involving God.

When I was a few years old, a woman held me in her arms and declared that she was my second mother. I was momentarily perplexed and a little embarrassed. But she soon cleared the air by saying that she had helped my mother deliver me. She was a village midwife. It was a custom in those days, at least in the village where I was born, that the midwives were called second mothers of the babies they helped deliver.

I looked closely at my second mother's face, and it indeed beamed with motherly looks. Then I asked her about how I was born. She explained the birth process with the utmost modesty. There was no mention of my father's part nor was there any attempt to explain how babies are conceived, not even how the birds and bees do it. She said it in plain words: "God created you in the womb of your mother."

Right after my birth (or as soon as I was cleaned up, I assume), someone recited verses from the Qur'an, piping the words directly into my ears: "There is no god but God...." The words, obviously, did not imprint on my memory, since I recall nothing of the event.

But I do understand the symbolism behind the ritual—the first sound a child should hear outside the womb must be the word God. In a way it is a literal enactment of the opening words from the Gospel of John, chapter one verse one, as quoted here from the King James version: "In the beginning was the Word ... and the Word was God." Thus my life began with the word "God," even though I could not hear it or understand it.

My parents did not preach religion to me. I learned it through living with them and being part of the culture that practiced it. My parents and my culture gave me my religion.

I was born in a small village where everything was small—a small school, a small mosque, and a small madrassa (religious school). I started school and madrassa at the age of five. I did not learn a lot about my religion at the madrassa. The class was held early in the morning (before sunrise). It started with kissing the Qur'an and reciting the text loud enough so that the teacher could hear us clearly. The text was in Arabic, which was unintelligible to all the students as well as the teacher. The whole mission of the class was to learn how to read and recite the Qur'an in a melodious chant. We were told that the words of the Qur'an are the words of God. We would learn the meanings of the script later in public school, but the recitation of the original words of God (relayed to the prophet Muhammad through the angel Gabriel) was an act of piety, a form of prayer. I believed in the sacredness of the Holy Book and its author, God. The only thing I did not like in the class was the teacher's strict discipline. He kept order in the class with a stick that he used to hit you hard if you couldn't read or if you mispronounced a word.

Memorizing the Qur'an was considered to be a great accomplishment as well as a great virtue and a noble tradition (the Qur'an was revealed to Muhammad in parts and was recorded as well as memorized by his disciples). However, understanding the meaning of its text was relegated to later times, mostly in schools where Arabic was taught as one of the languages. Translations of the Qur'an in the local language were available, but most of them were literal and would not make much sense without the guidance of a Qur'anic scholar. Thus, I learned to read the Qur'an but without understanding, and I could not memorize it except for a few passages. Nevertheless, I loved to recite the Qur'an or hear it recited by a professional. The melody is so sweet, awe inspiring, and soothing to the ears!

Although I was introduced to God very early in my childhood through everyday living and the reading of the Qur'an, the process wasn't like a theological study or indoctrination. The belief in God came as natural to me as my name. My parents gave me both my religion and my name, and they didn't ask me if I agreed or had any objections to either. As a matter of fact, most people inherit their religion, or a lack thereof, from their parents, and most spend their lives practicing what they inherited. Early childhood conditioning is the source of many of our beliefs, opinions, or prejudices that we continue to hold in our adult life. Although one is free to reexamine the past and make a few adjustments as we grow older, it is a hard nut to crack indeed! Old habits, beliefs, and prejudices don't change easily.

Invoking the name of Allah (Arabic name for God) in daily life became a second nature to all of us living in that culture. The words *Insha Allah* (God willing), *Shukar Al Hamdulillah* (thanks and praise be to God), *Bismillah* (begin with the name of God), *Allah O' Akbar* (God is great) and *Khuda Hafiz* (may God protect) were used in ordinary conversation as expressions of accepted truth, affinity, and goodwill. The word *God* somehow made the conversation smoother and more pleasant as it commanded veneration, trust, and commonality of faith. We began formal occasions with the name of God or with the recitation of a passage from the Qur'an to invoke God's blessing to the meeting. So in the village where I grew up, God was an integral part of my life. In a Muslim culture, we were constantly reminded of God through greetings, casual conversations, and formal prayers. We heard the call to prayers from the minarets five times a day. Some people were more religious than others, but none escaped the words, the sounds, or the thoughts of God in day-to-day living.

While living in a village during my childhood, I had the benefit of looking at the star-studded sky at night. I marveled at its beauty and mystery. I often wondered who had created such a beautiful dome over the world. The answer was automatic—God, of course—praise be to him. On many occasions, I saw shooting stars. What caused that to happen? The elders told me that it was a sign that something bad would happen to the world, so we had to be fearful of God's wrath and ask forgiveness. But we were never able to link a shooting star with any specific harm or calamity that followed its occurrence. There were skeptics, of course, who told us that it was nothing but a myth, an old wives' tale.

Besides worrying about shooting stars, there were unpleasant things that we had to observe now and then, such as the slaughtering of animals for food. As a child, I was awed by the sight of a chicken or a cow being slaughtered. But one thing that took the revulsion away from it was the fact that the slaughterer or the butcher first invoked the name of God before using the knife. That made it more humane.

There were occasions when I, alone or with others, prayed to God directly for help or guidance. One year, there was a terrible drought. All the streams and irrigation canals went dry. The crops were facing certain ruination. Each day we looked at the sky but there were no clouds to be seen in any direction. One day the word got around in the village that the people were going to appeal to God directly for rain. A large crowd, children and adults, gathered outside the village in a dried out, parched field. It was about noon and the sun was blazing high in the sky. Hundreds of us raised our hands as if in prayers while gazing at a few scattered clouds in the sky. Loudly, and in unison, we begged God for mercy and rain. I felt for the first time in my life that we were confronting God for his favor, face-to-face.

My recollection is that it wasn't long before we heard thunder and saw clouds heading in our direction. The sky turned dark and the sun hid behind the clouds. After a few loud thunders overhead, it started to rain and then it poured. The crowd went wild. I saw people crying, prostrating, or looking upward as if the real God was hiding behind the clouds. The torrential rains kept coming until they brought an end to the long drought. This event was talked about for many years. You may call it a miracle or a coincidence, but for me it was my first memorable introduction to God—a moment of truth that God was real and that he listened.

Since my first dramatic introduction to God, I have witnessed many events in which God's help was sought—from serious matters involving life and death to trivial concerns like victory for a sports team. Depending on the outcome, some results were acclaimed as miracles while others simply as God's will. However, no one had proof that God intervened in any case.

In the history of all religions, there have been incidents that were called miracles and were taken by the believers as proof of God's powers. But, history has also documented magical tricks and deceptions practiced by some who called them God's miracles. So it seems the miracles are just as mysterious as God himself and cannot be relied on as proof of God's powers or his existence.

Those who have studied both religion and science would agree that religion is based on faith and not on science. From the same token, they would acknowledge that science is the antithesis of faith. Many have tried to reconcile religion with science but have failed because science and religion don't mix. However, it is possible that a faith-based fact is corroborated by science. Without such evidence, it is either a miracle or a coincidence.

Recent studies in cosmology (a branch of science dealing with the origin of the universe) have brought science face-to-face with religion: who created the universe? The answer to this question may or may not turn out to be a supreme being or God as the creator, but we have never been closer to addressing this question scientifically as now.

I realize that the question of God or supreme being is not within the realm of science. No one can prove the existence of God through scientific methods. Cosmologists deal with the question of how the universe was created, not who created the universe. However, if we get close enough to answering the first question, it makes us reflect on the other.

Although it is easy to accept God as a matter of faith, it is not so easy to live up to his precepts. Acceptance of God not only entails belief in his existence but also in his attributes and code of human life communicated through prophets and scriptures. Some religions emphasize faith through deeds following God's edicts while others stress grace through worship and prayer. In either case, the path to God is tortuous. Believing in him is the easiest part.

God is invisible, and he has not revealed himself to mankind in a physical sense. He is perceived as a spirit or a force that created the universe and rules it through his infinite power and knowledge. Since science does not recognize anything outside of the physical realm, it cannot shed any light on a supernatural concept such as that of God. However, if it can concede the possibility of an agent, other than the natural forces or physical laws, responsible for creating the universe, the wall of separation between science and religion would crumble. Until then, the concept of God will remain as a matter of faith irrespective of whether or not a particular religious fact is corroborated by science.

This book is not an attempt to prove or disprove the existence of God. Rather, it is a search, through scientific and religious study, of the possibility of a cause or an agent responsible for the creation of our universe. Believing in God as the creator is easy. Finding the truth about how the universe came into being is the challenge.

CHAPTER 2
CONSCIOUSNESS

A number of years ago, I was brought into a hospital for an appendectomy. In my preparation for the operation, I was given an intravenous injection of an anesthetic drug. As the drug began to take its effect, I was asked to count backward from one hundred, which I did until I became incoherent and passed out. The next thing I remember is seeing some blurred figures standing in front of me. The blurriness soon gave way to a clear focus, and I could recognize the figures as those of my wife and daughters. It didn't take long before I realized that I was in the recovery room and that my family was there to greet me. From their smiles, I could tell that the operation had gone well and that I was alive.

I have absolutely no recollection of what happened between the time I lost consciousness in the operating room and the time I saw the blurry images of my family in the recovery room. I have often reflected on the complete blackout I experienced during surgery. Could it be that my center of consciousness—the part of the brain responsible for my being conscious of my surroundings and myself—went on the blink through a chemical reaction brought about by the anesthetic drug? If so, that seems logical considering the cause and the effect. The consciousness phenomena is definitely a physical process that can be turned on or off through chemical intervention.

We also know that consciousness resides in the brain, mediated by neurons that respond with a lightning speed to the internal and external electrochemical stimuli. Through a network of billions of neurons, the electrochemical signals received from inside and outside the body are processed and interpreted by the brain.

Neurological experiments have shown that centers of thought, feelings, memory, and consciousness are located in the cortex, the intricate outer layer of the brain. These regions are responsible for our mental activity and collectively called our mind.

Theologians and philosophers have debated the source of human thought and emotions through the ages. Many have attributed the mind's activity to man's soul or spirit, as if the brain were infused with some invisible, unphysical power linked to God. But modern science has refuted these claims by showing that the mind is nothing but a product of the brain's activity.

Consciousness is a mental experience that includes self-awareness, feelings, spatial orientation, memory, language, and thought. These processes occur in the brain as it synthesizes information received from the surrounding environment, from inside the body, and from the brain itself. Incoming signals from the sensory organs (eyes, ears, olfactory or smell senses, taste, and touch) are processed in the cortex and directed to specialized areas of the brain. Sensory inputs from the left side of the body are processed by the right hemisphere of the brain and those from the right side report to the left hemisphere. The information is shared by the two hemispheres through corpus callosum, a large cable-like bundle of nerve fibers. The brain transforms the electrochemical signals and gives rise to consciousness of both the inner and outer worlds. The region of the brain, called the association cortex, receives information only from within the brain itself and appears to be responsible for a person's awareness of his mental and physical state—the state of consciousness.

The terms *consciousness* and *self-consciousness* have been discussed extensively by philosophers, psychologists, neuroscientists, and other experts in the workings of the mind. Some consider the two as independent processes. For example, one can be conscious of one's surroundings while not being self-conscious and vice versa. Although the term self-consciousness connotes active perception of one's self as a unique individual, it is not independent of consciousness. In reality, the two processes are one and the same. The only difference is in the common usage of the terms—consciousness pertaining to awareness of one's surroundings and self-consciousness to one's self as a person. Other than this subtle distinction, one may consider consciousness and self-consciousness as two modes of the same process—the consciousness (or the self-consciousness).

We will use the terms consciousness, self-consciousness, awareness, and self-awareness to describe the same process—the brain's ability to interpret its internal and external environment. This interpretation by the brain is dependent largely on its past experiences, analyses, and perceptions stored in memory. The memory is instantly recalled to interpret the current experience.

Self-awareness or consciousness is an affirmation process that the brain goes through at any time to establish recognition of itself and its surroundings. It involves a very complex brain activity but the process is completely physical. Thoughts, feelings, action plans, and consciousness are states of mind brought about through the physiological activity of the brain.

Some religious philosophers believe that the brain controls the various bodily functions, whereas thoughts, feelings, and emotions come from God. It is obvious that such a belief has no scientific or physical basis. If there is a connection between God and the human mind, it is definitely beyond the scope of the human mind to understand.

Beginning of Consciousness

There is no scientific consensus as to when consciousness begins in humans. If we define consciousness as the ability to perceive the relationship between oneself and one's environment, then it is a common observation that this ability is not present until the baby is one to two years of age. However, if consciousness is considered simply as awareness of one's surroundings, then that ability could arise much earlier in life.

Consciousness can arise only when the brain is able to translate thoughts into nervous system commands, not just involuntary reactions to stimuli. This means that for consciousness to occur, the brain must be sufficiently developed and wired up to establish functional connections between the cortex and thalamus. Because these connections are not sufficiently established during gestation, fetal consciousness is ruled out on a physical basis. So it is reasonable to assume that human consciousness and self-awareness arise within the first two years of life.

Evolution of Consciousness

Consciousness is the product of brain activity. As the brain evolved in the animal kingdom, so did the faculty of consciousness. All animals with brains have the capacity for consciousness, but with varying degrees of its manifestation and characterization. For example, it is easy to recognize that dogs possess consciousness, but we do not know exactly what they are conscious of. Dogs have a keen awareness of their surroundings, as well as a well-developed sensory perception based on hearing and smell. Dogs can feel happy, sad, irritated, and afraid, and also have the ability to dream, but their capacity to think is limited. They do not show signs of reflective consciousness or any indication that they are "aware of being aware." For humans to have this ability is a great leap forward in the evolution of consciousness in the animal kingdom.

Is consciousness unique to humans? Many scientists believe it is not. Animals may not have the same awareness or experience the world as we do, but there is no reason to believe that they do not have inner experiences akin to some kind of consciousness. They are not able to verbalize their consciousness like humans, but many species have developed other ways of communicating their feelings or intentions. More research is needed to explore the inner world of animals and establish biological markers for its evolution. Similar exploration is also needed to measure consciousness in humans as it develops early in life.

Artificial Intelligence

The brain is often compared to a computer. Although there are similarities between computers and brains, there is a vast disparity between their capabilities. Whereas computers can be made to outperform any brain in solving computational and logical problems, there are fundamental difficulties in building an electronic machine that would work in exactly the same way as the brain. With the present state of scientific knowledge, one cannot say for sure if it is possible to build a machine that would mimic the human brain, especially in its seemingly unique functions such as consciousness, intuition, imagination, and emotion. Yet, it is intriguing to think that once we fully understand the workings of the human brain, we might be able to build a system that would be capable of generating thought and even self-consciousness. Current research in artificial intelligence is an effort in that direction.

Artificial intelligence (or AI for short) is the science of making machines, hardware, and software that are capable of making decisions, formulating action plans, generating thought, learning by themselves, and feeling emotion as humans do. Although AI is often thought of in the context of human intelligence, it is a broader science that includes humans as well as extra-human intelligence. In some respects, such as computation and data analysis, it far exceeds human intelligence. Computers outperform humans in tasks involving rigid logic and number crunching. In playing a chess game, they can beat the greatest chess master. Yet at things that we take for granted, such as consciousness, emotions, intuition, and common sense, even the most sophisticated computers fail miserably.

The idea that computers might someday have minds comparable to humans is a debatable issue. So is the notion that the human brain is simply a sophisticated computer. Nevertheless, there is no controversy among scientists that the human mind is the product of brain activity and that the mental experience is the result of interactions among billions of neurons within the brain.

The research in artificial intelligence is leading us in the direction of creating a sophisticated intelligent machine. Until that happens, any connection between the human mind and a God will continue to be debated.

PART 2

REASON

CHAPTER 3
SCIENTIFIC THOUGHT

The word *science* is derived from the Latin verb *scire*, meaning "to know." Webster's Dictionary defines science as "systematized knowledge derived from observation, study, and experimentation carried on in order to determine the nature or principles of what is being studied." [1] Scientific thought is the power of reasoning in accordance with the principles and methods of science. In the next five chapters, we will make use of scientific thought to explore the mysteries of our universe, its origin, its laws, and its evolution. The goal of this study is to see if science comes close to answering the question: how did the universe come into being?

Scientific thought is an inherent property of our brain to think and analyze our observations in order to find the true nature of what we observe. The hallmark of science is to seek the truth without prejudice. Although a complete lack of bias or prejudice is nonexistent among humans, given the nonscientific influences throughout their lives, the rules of science are clear, unambiguous, and unbiased.

In order to arrive at a scientific truth, one needs to follow the *scientific method*: (1) careful observation of the phenomenon; (2) tentative description or proposition, called a *hypothesis*, to provide a basis for further investigation; (3) experimentation to test the hypothesis and make modifications if needed; (4) formulation of a theory. A consistent hypothesis becomes a *theory*.

A theory is a tested formulation of relationships between observations and predictions—a framework in which the underlying principles of the observed phenomenon have been verified through experimentation or the application of the established scientific facts. When a hypothesis becomes a theory, it is implied that considerable scientific evidence exists in support of the formulated principles.

If the theory reveals a basic fact or truth about nature, it may be elevated to the status of a *law*. The law is a scientific truth that has stood the test of all observations and experimentations without exception.

Although a theory or a law may be formulated through the scientific process with utmost rigor, it cannot claim to be an everlasting truth. Its reign extends only to the time when a new theory or a new law comes along which proves the previous one to be wrong or less accurate. Nothing in science is considered absolute, permanent, or immutable. Science establishes a theory with the same rigor as it demolishes it. The same fate awaits a theory or a law as human knowledge progresses through millennia.

The fact that theories or laws can be revised or abandoned in face of new scientific evidence does not, however, affect the truth itself. The theory may change but the truth remains to be found. The object of science is to reach the truth, but in the process it leaves the door open for further investigation. A theory is the closest approximation to the truth at the time it is accepted by the scientific community.

Newton's law of gravity is a prime example of a theory that had been well established through centuries and yet was not immune to being changed, revised, and even discarded. About two and a half centuries after Isaac Newton proposed his theory of gravitation, Albert Einstein published his general theory of relativity, which changed the whole concept of gravity. Although Newton's law is still valid in most situations in our everyday life, it has shown slight discrepancies in some of the astronomical data. Einstein discarded the Newtonian view of gravity as a force between masses and replaced it with his concept of gravity as a distortion of space-time by mass. We will revisit gravity later, but make the point here that the scientific theories, by definition, are not immortal. Science evolves with the accumulation of knowledge, and so do its theories and laws.

Scientific thought has given rise to many classes of science with varying degrees of rigor and discipline. The mother of all sciences is mathematics because it confers on other sciences their rigor, exactness, accuracy, and precision. It is the principal tool of the human mind to figure out magnitudes, quantities, forms, and their relationships. Arithmetic, algebra, geometry, and calculus are some of the common mathematical tools that we use to analyze scientific problems and authenticate their solutions. A mathematical formula, or an equation, is the most succinct way of describing a scientific fact, a theory, or a law.

Next to mathematics, the most basic of all the sciences is physics. It deals with the physical properties of the universe, the laws that govern its being, and what is in it. In a sense, it is a science of everything—matter, energy, forces, space, time, and their interactions. Physical forces at the atomic and molecular level are responsible for what we observe in chemistry and biology. The deeper we go into the study of life, the closer we get to the physics that underlies the vital phenomena. We can explain a lot about our bodily functions, the working of our brain, the biology of living cells, and the genetic molecule, the DNA, but our knowledge of life is still limited to the cellular and molecular level. We know very little of how the factory of life operates at the electronic level. Therefore, without a full understanding of the physics of life, we cannot create it. That said, our knowledge of physics, chemistry, and biology has been advancing at an enormous pace during the last century. It is not inconceivable that the code of life will be broken down to its most basic level in this century, enabling humans to create life from its elementary ingredients.

There are many different fields of science, not all of which arrive at conclusions in the same way. Physical sciences (physics, chemistry, biology) use the most rigorous forms of the scientific method. They are called the exact sciences because they use mathematical tools to analyze and quantify their results. Relatively less mathematical are the descriptive sciences, like botany, zoology, and anthropology, which use descriptions and inferences to arrive at results. Yet, they also follow the rules of the scientific method, namely, observation, hypothesis, experimentation, theory, and law. Scientific thought, the power of scientific logic, permeates the whole process of scientific discovery and the validation of the scientific truth.

Is there a truth other than the scientific truth? It is possible, if the truth is taken as the absolute truth. Because science never claims absoluteness of a truth in spite of its goal to reach it, it does not rule out an alternative theory or concept. However, if the alternative theory or concept fails the scientific test (through scientific method), it is rejected. Thus, the acceptance or rejection of a fact as truth can only be accomplished through the scientific method. If a theory cannot be proved or disproved through the scientific method, it is considered unscientific or beyond the scope of science. Even then, science would not reject the theory without the scientific evidence to reject it. Science is a disciplined knowledge. It knows its strengths as well as limitations. True science does not exceed its scientific boundaries, nor does it stretch the truth.

Scientific thought can be speculative as long as it is labeled as speculative. Scientists often guess or speculate about things during a study, but the intent is to explore new ideas for investigation rather than get around the scientific method. It is always acknowledged that a theory based on speculation is not an established fact. But, like science fiction, a speculative theory may excite the imagination and even prompt further investigation.

Beware of the pseudoscience, however. A pseudoscience is a body of knowledge that presumes without warrant to have a scientific basis. It may be camouflaged as science with the use of a few superficial trappings of science, but it is not science in reality. It is a kind of masquerade of science.

Pseudoscientific theories are known to rely on anecdotes, ancient myths, legends, and intuition. Instead of following the scientific method for legitimacy, which involves careful observation and experimental evidence, they often use the gullibility of the believer as a tool to strengthen their base. For example, some of these theories are based on ancient myths, philosophies, and religious scriptures, but such theories do not involve the substance of science. Their evidence is contrived to confirm infallible dogmas rather than a proven fact of nature.

There are many categories and subcategories of science. Some are well established and fully recognized as science in that they follow the scientific method of inquiry to arrive at the truth. However, there are some that do not always adhere strictly to the scientific criteria. Philosophy is one of them. Philosophy may or may not be a science, depending on whether it uses scientific or speculative logic.

Philosophy, as a discipline, is defined in Webster's Dictionary as "theory or logical analysis of the principles underlying conduct, thought, knowledge, and the nature of the universe: included in philosophy are ethics, aesthetics, logic, epistemology, metaphysics, etc." [2] The definition, by itself, does not stipulate the conditions under which philosophy must operate, but its use of logic to explain observations and concepts, scientific or otherwise, is commendable.

Philosophy can be an effective analytical tool for many sciences that occasionally teeter on the brink of being a science or a belief. With its incisive analysis, it can separate the two. The scientific role of philosophy is to clarify and criticize the results. In a way, the method of philosophy is not very different from that of mathematics. Both are used to analyze the validity of the scientific method. However, unlike mathematics, philosophy is used to analyze the original premises of the theory rather than deriving quantitative relationships between its parameters.

There is another kind of philosophy called *speculative philosophy*. It uses logic to reflect on nonscientific facts, religious experiences, and intuition. Sometimes it is used to bring logic to what is generally considered as pseudoscience. In the process it becomes a pseudoscience itself.

In earlier times, philosophers considered science as part of philosophy because they could study and understand both. However, in modern times, science has become too technical and mathematically complex for the philosophers to follow. So, the trend now is for the scientists to learn philosophy and apply its tools of logic to justify their discoveries in addition to the scientific method. It is expected that scientists would use analytical rather than the speculative kind of philosophy to rationalize and explain their findings.

Analytical philosophy avoids the speculative aspects of philosophy as much as possible. It is an integral part of scientific thought that, in the final analysis, is the combination of scientific method and analytical philosophy. Thus, scientific thought is the philosophy of science and vice versa. It is man's power of reason and logic without bias or prejudice.

There is a class of scientific theories that lack sufficient experimental evidence to become well established. These may be called evolving theories. They are called scientific because they are based on other theories and laws that are well established. For example, Einstein's general theory of relativity is accepted by most scientists because it is based on the established laws of physics and mathematics, albeit very complex. Questions of its theoretical validity and inconsistencies have been addressed by scientists in the last ninety years and some experimental data have been gathered to confirm parts of this great theory. However, because of its complexity and the difficulty of experimental verification as a whole, it is still called a theory. This implies it needs further experimental verification before attaining the status of a law, although parts of it are well established and merit the status of laws.

There are many theories regarding the origin of our universe. Most of them are evolving theories. They are based on mathematical and physical principles, but not all of them are ironclad with respect to experimental evidence. Far from being established, their theoretical models are still under investigation by scores of theoretical and experimental scientists. Some theories are discarded almost as soon as they are proposed; some barely survive and continue to evolve. The reader should understand that for any set of ideas to become a viable scientific theory, it must meet the scrutiny of science. Scientific theories are not speculative beliefs. They must meet the scientific criteria to be called theories, namely, careful observations and experimental verification. Even then a theory is tentative until better theories come along and supersede it.

CHAPTER 4
LAWS OF PHYSICS

In the following chapters, I will discuss major theories and concepts that have been formulated by scientists to explain the origin of our universe. However, before I do that I would like to familiarize the reader with some basic physical principles that underlie these theories. This is a vast subject, but the intent here is to help the general reader understand the highlights of what we know about our universe and how it began. However, if you, the reader, are not particularly interested in the basic physics underlying the cosmos, feel free to skip the physics jargon wherever it occurs and just concentrate on the concluding facts of the theories.

It is assumed that the reader is at least somewhat familiar with high school or beginning college level physics and can recollect, even though vaguely, Newton's laws of motion and gravity: the first law regarding inertia, the second about force and acceleration, and third law equating action to its opposite reaction. Newton's law of gravitation defines gravity as a force between objects that is proportional to the product of their masses and inversely proportional to the square of the distance between them. That means that gravitational force between objects is greater if they are more massive and the distance between them is shorter. Newton's law of gravitation gives the mathematical relationship between the gravitational force, the object masses, and the distance between them.

In Newtonian physics, mass and energy are distinct entities, quantities that can be measured when the objects are moved against inertia or accelerated. If a force acting on an object gives it an acceleration, then the magnitude of the force is given by the mass of the object times the acceleration. In other words, equal masses offer the same resistance to acceleration. From the same token, larger masses require a larger force proportionally to produce the same acceleration.

We commonly refer to mass as weight but the latter is not the same quantity as mass. The weight of an object is the force with which the earth attracts it. Mathematically, weight equals mass times the acceleration due to gravity.

It can be shown from the Newton's law of gravitation that objects dropped from the same height above the earth will experience the same acceleration irrespective of their masses and, if there were no resistance caused by the intervening air, they would take the same amount of time to fall to the ground. Galileo (1564–1642) verified this result experimentally when he dropped different weight objects from the Eiffel Tower through evacuated tubes (to eliminate air resistance). This experiment showed that the acceleration due to Earth's gravity is constant at a given elevation (about 32 feet per second squared close to the earth's surface) and does not depend on how heavy the falling object is. Gravitational acceleration is determined predominantly by the massive earth and not by the falling object whose mass is negligibly small compared to the mass of the earth. However, more massive objects accelerated by gravity have greater force and greater momentum.

If a force acting on an object moves it through a distance, it is said to have performed work. Capacity to do work is called energy. An object moving with a velocity has momentum and energy. This form of energy due to motion is called kinetic energy.

Another form of energy, called potential energy, is the energy of an object due to its position. For example, if an object on the ground is raised to a higher level such as on top of a building, it is said to have acquired a potential energy. This energy can be retrieved if the object is allowed to fall back to the ground. If the potential energy of an object located far away from Earth and outside Earth's gravitational influence is arbitrarily set to zero, then its potential energy at a point where it nears the earth decreases below zero. Its potential energy due to gravity is denoted as negative. The closer the object is to the earth's center, the more negative its potential energy. On the other hand, if we pull the object away from the earth, we add energy, thereby increasing its potential energy or making it less negative.

$E = mc^2$

It is a well-known fact that mass and energy are interconvertible. Einstein's famous equation, $E = mc^2$, where E is the energy, m is the mass at rest, and c is the velocity of light, quantitatively describes the relationship between matter and energy. The energy released in an atomic bomb explosion is one example of a small amount of mass being converted into a vast amount of energy in the form of explosion, heat, light, and radioactivity.

Radioactivity is the emission of particle and photon radiation. Photons are called electromagnetic radiation because they travel through space by the mechanism of oscillating electric and magnetic fields associated with photons. This mode of energy propagation characterizes such phenomenon as light waves, heat waves, radio waves, ultraviolet rays, X-rays, and gamma rays. James Maxwell (1860s) called these radiations "electromagnetic" and described them as oscillating electric and magnetic fields. The electric field and the magnetic field associated with photons radiate at right angles to each other and at right angles to the direction of motion of the photon at any instant of time. The energy carried by these oscillations is propagated in space with the maximum possible speed, the speed of light (3×10^8 meters per second or 186,000 miles per second in vacuum). The phenomenon of energy propagation by oscillating fields is analogous to energy traveling through the action of water waves.

Einstein published his special theory of relativity in 1905 and discovered the phenomena called photoelectric effect and Brownian motion. He explained the photoelectric effect—a process in which a photon is absorbed by an atom, thereby causing the ejection of an electron from its orbit—by theorizing that the electromagnetic radiation is propagated in space as discrete packets of energy rather than continuous waves. The quantum nature of photons—an energy packet that can collide with material particles as if it was a particle itself—has been demonstrated by observing photon interactions with matter such as the well-known phenomena of Compton effect and pair production.

The concept of quantum or particle nature of electromagnetic radiation does not do away with its wave properties. The latter can be demonstrated by experiments involving phenomena such as wave interference and diffraction in crystals. Thus, light or any electromagnetic radiation has a dual nature. When its frequency of electromagnetic oscillation is low (or its wavelength is large), it acts more like a wave, although still being a photon with a small but discrete amount of energy. As the frequency of oscillation increases (e.g., in the range of x-rays and gamma rays), the electromagnetic oscillations are so compressed together that the photon acts more like a particle when it passes through matter and interacts with atoms. It can be absorbed by atoms to cause ejection of electrons (photoelectric effect) or collide with an electron in a manner analogous to a billiard ball collision (Compton effect).

A photon is characterized by its frequency—the number of oscillation cycles per second or hertz—or its wavelength—the distance between crests or troughs of the electromagnetic waves. Its energy, frequency, and wavelength are related mathematically.

The spectrum of electromagnetic radiation in nature is quite wide with wavelengths ranging anywhere from 10^7 meters (radio waves) to 10^{-13} meters (ultra-high frequency or high-energy X-rays). Since wavelength and frequency are inversely related, the frequency spectrum corresponding to the above range is anywhere between 10 and 10^{21} hertz. Energy is directly proportional to frequency, so the energy range of the above spectrum would be 10^{-13} (very low for radio waves) to 10^7 (very high for energetic X-rays) as expressed in units of electron volts.

The Atom

All material objects are composed of individual entities called elements. Each element is distinguishable from the others by the physical and chemical properties of its basic component—the atom. Each atom consists of a small but very dense central core, called the nucleus, where most of the atomic mass is located, and a surrounding "cloud" of electrons moving in orbits around the nucleus. Whereas the radius of the atom (radius of electronic orbits) is about 10^{-7} millimeters, the nucleus has a much smaller radius, about 10^{-11} millimeters. Since electrons have an extremely small mass and move in orbits of diameter ten thousand times larger than the nuclear diameter, the atom is mostly an empty space with its mass concentrated almost entirely at the center of its nucleus.

The properties of atoms are derived from the constitution of their nuclei and the number and organization of their orbital electrons. The nucleus has a structure of its own and contains two kinds of particles, protons and neutrons. Whereas protons are positively charged, neutrons have no charge. Since the electron has a negative charge and the proton has an equal amount of positive charge, the number of protons in the nucleus is equal to the number of electrons outside the nucleus, thus making the atom as a whole electrically neutral.

The lightest atom found in nature is that of hydrogen. It consists of a single proton as its nucleus with an electron orbiting around it. The helium nucleus has two protons and two neutrons, with two electrons revolving outside in the atomic orbits. The oxygen atom has eight protons and eight neutrons in its nucleus and eight electrons moving in their atomic orbits and so on. For some elements the number of protons equals the number of neutrons, while in others, especially with heavier atoms, the number of neutrons exceeds the number of protons.

There are a total of 103 elements known today. Of these, the first 92 (with proton number equaling 1 to 92) occur naturally. The others have been produced artificially (e.g., in reactors). In general, the elements with less number of protons in the nucleus are stable and the ones with greater number of protons are radioactive. It appears that as the number of particles inside the nucleus increases, the forces that keep the particles together become less effective in keeping them together, and, therefore, the nucleus becomes unstable or radioactive, and emits radiation in the form of particles. The particle emission from a radioactive nucleus is often accompanied by gamma ray photons in order for the nucleus to get rid of its excess energy.

Atoms may emit radiation (particles or photons) when subjected to internal or external forces. Electrons can change orbits or get captured by nuclei, resulting in X-ray emission. A high-speed electron passing by a nucleus may be attracted so hard by a nucleus that it may lose all or part of its kinetic energy. The energy in this case appears as X-ray photons. Nuclear transformations or disintegrations give rise to radioactivity. So at the atomic level, matter and energy can be inter-converted through a natural process such as radioactivity—or artificially by bombarding atoms and nuclei by energetic particles.

The process of fission, in which nuclei are made to split in a sustained nuclear reaction, converts a part of nuclear mass into a vast amount of energy in accordance with the $E = mc^2$ equation. In another process called fusion, hydrogen nuclei can be made to fuse together under intense heat conditions such as those existing in our sun and other stars. Some of the nuclear mass in this process is converted into energy. The stars are nuclear furnaces that are continuously fusing nuclei together and thereby converting matter into energy.

Electrons are particles with negative charge. Positrons are the same particles, namely electrons, but with positive charge. This is an example of *matter* having its corresponding *antimatter*. Electrons and positrons can combine to annihilate each other. Their combined mass disappears and is converted into energy. This energy is emitted as two photons radiating out in opposite directions. The reverse—creation of matter from energy—is also possible. A high-energy photon, passing by a nucleus of an atom, may completely disappear. The energy of the photon is converted into matter—an electron-positron pair of particles. Part of the photon energy is used up in creating mass, (i.e., the electron-positron pair). The rest of its energy is shared between the particle pair as kinetic energy, making them race away in different directions from the site of their creation.

In the above examples of mass-energy interconversion, the total energy (considering mass as a form of energy) remains constant. In physics we call it *the law of conservation of energy*.

Our universe is made up of matter and energy. Although the universe is expanding, the stars are burning, and mass and energy are being interconverted at the atomic level, there is no net loss or gain of energy of the universe. No new energy is created without destroying mass, and no new mass is created without expending energy.

A question arises as to what is the total energy of the universe. The answer is that it is zero. All the matter, antimatter, and photons in the universe represent a positive energy, while the gravitational field has a negative energy.[1] The positive energy exactly balances the negative gravitational energy. *The total energy of the universe is zero.*

Forces of Nature

At the very start of the universe, there was probably one force that governed its creation and survival in the initial 10^{-36} seconds of its existence. The single force then broke down into four separate forces at the end of 10^{-10} seconds. Since that time, the universe has managed itself with these four forces.

The four forces of nature today are, in order of their strengths, (a) strong nuclear force, (b) electromagnetic force, (c) weak nuclear force, and (d) gravitational force.

Of these forces, the strong nuclear force is the strongest, but it is a very short-range force. It comes into play when the distance between particles becomes smaller than the nuclear diameter (about 10^{-11} millimeters). This force plays a part in forming heavier nuclei when nuclei are brought very close together. It is also responsible for keeping protons and neutrons together inside the nucleus and preventing them from escaping. However, if sufficient energy is provided to any of the particles inside the nucleus, it can overcome the "barrier" provided by the strong nuclear force and escape. This is what happens when a nucleus becomes radioactive. A naturally occurring radioactive nucleus has excess energy (due to being packed with too many protons and neutrons or undergoing a transformation in which a neutron converts into a proton and vice versa) and allows a particle to escape. Excess energy can be provided to the nucleus also by bombarding it externally with particles. If the bombarding particle gets close enough to the nucleus, strong nuclear force will pull it in, thereby making the nucleus heavier and more unstable or radioactive.

As mentioned earlier, nuclei can fuse together. If they are brought close enough for the short-range strong nuclear force to overcome the nuclear barrier they could be pulled in to form one fused nucleus. This is what happens when nuclei are subjected to extreme temperatures such as those prevailing in our sun and the stars.

The electromagnetic force (it is so called because electricity and magnetism are fundamentally the same by nature) causes oppositely charged particles to attract each other and similarly charged ones to repel each other. Although this force tends to disrupt the nucleus containing protons because it makes protons repel each other, the strong nuclear force, being much more powerful, overpowers the electromagnetic repulsive force and keeps the nucleus intact.

The weak nuclear force is quite weak, as its name suggests. It plays a part in certain types of radioactivity emissions (e.g., beta-particle decay). Because the gravitational force is the weakest of all the natural forces, it does not play any part at the nuclear or atomic level. However, it has played a huge part in the creation and evolution of our universe.

Elementary Particles

In the early ages, before the Common Era, Greek philosophers thought about the basic constituents of matter. Plato maintained that one could never understand the universe without knowing the smallest component of matter. Democritus called the smallest entity of matter, which could not be further subdivided, an atom (in Greek, *atomos*; *a*, meaning "not," and *tomos*, "to cut"). He believed that there was nothing except atoms and empty space.

For centuries after Democritus, the atom was considered to be the smallest and an indivisible particle of matter. Chemistry thrived with the concept of the atom, as all elements were characterized by the physical and chemical properties of their atoms. After the discovery of electron by J. J. Thomson in 1898 and the introduction of Niels Bohr's theory of atomic structure in 1913, the atom became the "miniature universe" for the physicists, to be explored by some of the greatest minds of the twentieth century. Not content with discovering the structure of the atom, with its orbital electrons and central nucleus packed with protons and neutrons, physicists continued this search for the original "a-tom" of Democritus—the indivisible particle of matter. Many such particles have been discovered. They are called the elementary or fundamental particles. These particles may be considered the most basic constituents of matter—indivisible, structureless particles of negligibly small size.

There are two kinds of fundamental particles of matter: quark and lepton. There are six types of each of these, as listed below:

Quarks: up, down, charm, strange, top, and bottom;
Leptons: electron, electron neutrino, muon, muon neutrino, tau, and tau neutrino

Besides the above twelve basic particles of matter, there are twelve corresponding basic particles of antimatter. This follows the principle discovered in 1928 by Paul Dirac (1902-1984), which states that for every particle of matter there must be another particle of antimatter with the same mass but opposite charge, so there are six antiquarks and six antileptons.

Quarks are the building blocks of heavier particles such as neutrons, protons, and mesons. For example, it takes three quarks to make a proton. These quarks are held together by force-field particles called *gluons*, the messenger particles of the strong nuclear force.

According to the quantum electrodynamic (QED) theory, messenger particles are the carriers of force in a force field. These particles of force are not material particles but quanta (packets of energy) of the field like the photon. Thus, the force between any two interacting matter particles is transmitted by the messenger particles, the quanta of the field, traveling at the speed of light, which is the speed with which all photons travel.

It may be mentioned that basic particles are characterized not only by their masses or quanta but also by their spin. For example, the matter particles (quarks and leptons) have noninteger spin (e.g., 1/2, 3/2) and are collectively called fermions. The messenger particles, on the other hand, have integer spin (e.g., 0, 1, and 2) and are collectively called *bosons*.

There are thirteen messenger particles or bosons that mediate the four forces of nature. They are listed below:

Electromagnetism	—Photon (γ)
Strong force	—eight gluons
Weak force	—W^+, W^-, Z^0
Gravity	—graviton (not yet detected)

Whereas matter particles or fermions can attain high energies or speeds, they cannot quite attain the speed of light. When their speed reaches close to that of light, further acceleration increases their energy through an increase in their mass rather than their speed. So the ultra-high-energy particles produced in accelerators (e.g., Tevatron at Fermilab and a giant atom smasher at Européèn pour la Recherche Nuclaire (CERN) in Geneva) have greater mass but are not as swift as light. The bosons, on the other hand, can have high quantum energies, but they all travel with the speed of light. They can also transform themselves into material particles, whereby their high energy is converted into high-energy material particles.

Another mysterious particle so far undetected has been added to the above list. It is called the *Higgs boson*, after Peter Higgs, who predicted its existence in 1964. Particle physicists postulate that our universe is pervaded with Higgs bosons. The Higgs field is thought to permeate all space and is the same everywhere. All the mass possessed by matter is generated by the Higgs field pervading the universe. In other words, particles acquire their mass through interaction with the Higgs field. The sea of Higgs bosons produces a drag effect on the particles, thereby manifesting properties of inertia. The resistance to motion in the sea of Higgs bosons defines their mass.

It should be mentioned that Higgs's idea was used by theoretical physicists Steven Weinberg and Abdus Salam to combine electromagnetic and weak forces into a unified electroweak force, mediated by messenger particles, photon, W^+, W^-, and Z^o.

Back to the Higgs field: the term *field* in physics is defined as "lines of force." For example, a magnet is surrounded by its magnetic field. A particle of iron placed in the field would be attracted toward the magnetic pole and follow a path or a line of magnetic force. Henning Genz has a more imaginative description of a field: "an excitation of space that oscillates like tall grasses in a heavy wind."[2] The forces are transmitted in a field by the exchange of force carriers such as photons, W or Z bosons, and gluons. For the Higgs field, the force carrier is the Higgs particle (a boson).

We do not have experimental evidence to confirm the existence of Higgs field or its boson, the messenger particle. But an experiment with CERN's Large Electron-Positron collider recently produced a shower of particles with characteristics similar to those hypothesized in Higgs boson decay.

The Uncertainty Principle

One of the most bizarre predictions of the quantum theory is a law known as the *uncertainty principle*. It was discovered in 1927 by Werner Heisenberg (1901–1976). It states that we cannot simultaneously determine the position and speed of a subatomic particle. In other words, measuring one quantity diminishes the precision with which we can measure the other. No one understands why this principle exists, but its predictions agree well with the experimental results. It is an empirical fact but it has acquired the status of a fundamental law of nature.

To understand the uncertainty principle a little more, let us attempt to measure the position and velocity of an electron. It is a well-known fact in physics that all material particles are not only particles of mass but they also exhibit a wave nature. This dual nature of matter was discovered by Louis-Victor de Broglie (1892–1987), who proposed the concept in his PhD thesis in 1924 at the University of Paris (de Broglie is the only physicist who won the Nobel Prize on the basis of a PhD dissertation). So, an electron is not only a material particle, it is associated with a *de Broglie Wave*. Because of its wave nature, the particle's position is uncertain. In accordance with the quantum mechanical principles, if we measure the position or velocity of an electron, we will get a probability distribution. The Heisenberg uncertainty principle predicts that if the spatial distribution (to pinpoint its location) is narrow, its velocity distribution will be broad. Conversely, if the velocity is precisely known (narrow velocity distribution), its location will be more uncertain (broad spatial distribution).

Quantum mechanics and its uncertainty principle are believed to be implicated in the creation of our universe. We will discuss their role in the cosmic origin in the next chapter.

Classical Gravity

Newton's laws of motion and gravity work very well when objects are moving at speeds much less than the speed of light and where gravity is relatively weak, such as in our solar system. But the accuracy of Newton's laws is reduced significantly when the speed of objects gets close to that of light and the gravitational field gets very strong, as exerted by massive stars, quasars, galaxies, and black holes. The concept of gravity as a Newtonian force breaks down under extreme conditions of gravitational field.

In 1916, Albert Einstein published his general theory of relativity, in which he described gravity as a warping of space-time around a massive object. This was a revolutionary leap in our understanding of space, time, and gravity. He constructed a complex set of mathematical equations that predicted that an object moving in a gravitational field would follow curved trajectories dictated by the curvature of space and time caused by gravity. The often-used analogy is to consider space as the surface of a trampoline. Place a heavy object on its surface, and it will create a dent. Now, if you roll a ball across its surface, it will move in a straight line but follow a curved path because of the dent.

It's a common observation that long-distance flights between cities follow curved paths. The plane's flight path looks curved on a flat map, but it is the shortest distance between the two cities as the plane flies parallel to the curved surface of the earth. A similar situation exists when light passes by an object like a massive star. It is deflected toward the object because it travels along the curved space (caused by intense gravitational field), taking the shortest path between two points.

The greater the mass of the object, the stronger is the gravitational field, the greater is the curvature of space and time, and the larger is the deflection of light's path. But it should be realized that objects, including light, take the shortest distance between two points, and if the space is curved due to gravity, their path is still the shortest along the curved space.

Whereas Newtonian gravity laws fail in intense gravitational fields, Einstein's general relativity equations predict accurately the motion of objects in such fields. However, Einstein's theory does not explain the true nature of gravity. It says that it is not exactly a Newtonian force but somehow it distorts space and time.

Mathematically, the effect of gravity manifests itself as a distortion of space and time together. Einstein called it space-time, as if space and time were linked together in a geometry of four dimensions—three for space and one for time.

Although in our ordinary experience, space and time are separate entities, an object moving in a gravitational field experiences distortion of space and time simultaneously as if space and time were wrapped together into one geometry with four dimensions: length, width, thickness, and time. Any event can be located with three coordinates in space and one in time.

Einstein first envisaged the linkage between space and time in his special theory of relativity (1905). He theorized that the fundamental laws of physics do not depend on location or motion. For example, the speed of light would be measured the same by any two observers regardless of their speed relative to each other. Suppose an observer A is moving at half the speed of light toward another observer B at rest. Let us assume that both observers have equipment to measure distance (e.g., a ruler) and time (e.g., a clock). Observer A sends out a signal of light in the direction of observer B. Both observers measure the distance and the time that light travels between two points to measure the speed of light (distance/time). They will both measure the same speed of light (186,000 miles per second), although the Newtonian law of motion would predict that observer A should measure 1.5 times the speed of light compared to observer B. The reason that both the observers measure the same speed of light is that the distance and time measured by the two observers are different because of their relative speeds. Observer A will measure a longer distance (shorter ruler) and a longer time (slower clock) while observer B will measure a shorter distance (larger ruler) and a shorter time (faster clock). But they will compute the same speed of light, which is the ratio of distance over time.

The special theory of relativity predicts that objects moving at relativistic speeds close to the speed of light will get shorter in dimension along the direction of their velocity and their clock will run slower (less time) compared to an object at rest. The ratio of distance to time, which is speed, still remains the same. These phenomena have been experimentally verified and point toward a close linkage between space and time. Einstein lumped space and time together and called it space-time when describing the distortions caused by gravity with regard to space and time. In his general theory of relativity, Einstein related the curvature of space-time with the distribution of mass. The Newtonian force of gravity was replaced by distortion created in the space-time matrix by mass and energy.

A gravitating body, like a massive star, curves the surrounding space as well as slows down time. Thus, the physical reality of gravity is represented by a field with four independent variables—the three coordinates of space and one of time. The theory makes accurate predictions with regard to bending of light in an intense gravitational field but still needs direct experimental verification of this phenomenon being caused by distortion of space-time.

In 2004, NASA (National Aeronautics and Space Administration) launched a spacecraft carrying Gravity Probe B to test two predictions of Einstein's general theory of relativity: warping of space and time by the presence of Earth and dragging of space-time as it rotates. The spacecraft is equipped with very sensitive gyroscopes to precisely measure tiny changes in their spin as the satellite orbits the earth at 400 miles altitude directly over the poles. The results are awaited pending the successful completion of the experiment.

Einstein's theory also predicts that accelerated masses, such as binary stars moving in orbits, give rise to gravitational waves in space-time. A gravitational wave expands and contracts space-time. These undulations of space-time are difficult to measure, but a number of experiments to detect gravitational waves are under development.

Quantum Gravity

Of all the forces of nature, gravity is the most familiar but the least understood phenomenon. Einstein's description of gravity was an improvement on the Newtonian concept of gravity, but it is still considered to be a classical theory in the context of modern physics. The notion that gravitational forces result from the distortion of space-time by mass and energy still cannot explain the nature of this force, especially under conditions that prevailed in the early universe. Other forces, namely, the electromagnetic, the strong nuclear, and the weak nuclear, have been unified by the modern theory of quantum mechanics. These forces are mediated by the bosons that travel in the field associated with particles at the speed of light. But a similar theory for gravity has not yet been developed. Einstein's theory of gravity, the theory of general relativity, seems to be incompatible with the theory of quantum mechanics.

However, this does not prevent scientists from hoping that a quantum theory of gravity could be developed in which the gravitational force is carried by the exchange of messenger particles, the "gravitons." So far, the effort to merge general relativity into a quantum field theory has not been successful.

Another approach that is showing promise for unifying the theory of forces is the idea of replacing point-like particles (fermions and bosons) with infinitesimal *strings*. A theory of strings that would explain all the forces of nature including gravity is still in its early stages of development. We will discuss its prospects in the next section.

Theory of Everything

In order to understand our universe, we need to understand the forces that govern it. At present there are two theories that partially explain the four fundamental forces of nature: the quantum theory and the general theory of relativity, as discussed earlier. The fourth force, gravity, may involve a messenger particle, graviton, but its existence has not been verified experimentally and therefore it cannot be explained by the quantum theory.

One problem in the above dilemma is that the quantum theory and the general theory of relativity are diametrically opposed to each other. Whereas, the former is the theory of the microcosm dealing essentially with the subatomic world, the latter is a theory of the macrocosm, the world of massive stars, galaxies, clusters, and black holes. Although the macrocosm is the aggregate of the microcosm, neither of the theories is general or fundamental enough to explain both in a unified framework.

Scientists have struggled for over fifty years to come up with a unified theory of the forces that were responsible for the creation and evolution of our universe. All attempts to merge the quantum theory with the general theory of relativity have failed. So the time has come to say farewell to these partial theories and embrace a radically different idea. This idea or theory, which is currently in its infancy, is called the *superstring theory*. It is purely a mathematical formulation with very little possibility at the present time to verify its validity experimentally. Accordingly, it has raised a storm of controversy in the scientific community.

The superstring theory postulates that the fundamental constituent of matter and energy is an infinitesimally small string of energy vibrating in ten dimensions. These strings can be open, like hair, or closed like loops. Vibration of the elementary strings of energy and their interactions with each other create different frequencies. Each resonant frequency can give rise to a subatomic particle or a quantum of energy.

The superstring formalism has shown that the string vibrations cause spacetime to be distorted, as predicted by Einstein. Thus, the superstring theory effectively merges the quantum theory and the general theory of relativity into one theory. If developed to full maturity and verified experimentally, it will be called the "theory of everything" or TOE.

There seems to be an abundance of string theories describing matter and forces that make up the universe. The most prominent ones are five: three of superstrings and two of heterotic strings. It turns out that these five string theories are like an "island on the same planet." In other words, they represent different aspects of the one underlying theory, called the *M-theory*. M may stand for mother or mystery, but it is supposed to be the theory of everything.

The M-theory has a membrane, as opposed to a string, as the fundamental constituent of matter and energy. Consequently, it lives in eleven dimensions rather than ten. As mentioned earlier, these dimensions provide degrees of freedom for the strings or membranes to vibrate. Some modes of vibration, or "notes," give rise to fermions (quarks and leptons) while others appear as bosons (photons, gluons, W's, Z, and gravitons).

An explanation is needed at this point for the concept of the extra dimensions. We all recognize three dimensions of space and one of time. What about the additional six dimensions for the string theory or seven for the M-theory? Although one does not experience ten or eleven dimensions of space in real life, we can assume that the extra dimensions are curled up very tightly, beyond our recognition. We are not aware of their existence, but mathematically we need to assign extra dimensions or degrees of freedom in order to characterize different particles and bosons by their mass, charge, and spin. The string theories point to extra dimensions of space at the time the universe came into being and later when, through phase changes, it settled down to what we have today—three of space and one of time.

In summary, various string theories are being developed to explain on a more fundamental level (1) the nature of four forces, (2) the existence of elementary particles, (3) the origin of matter and energy, and (4) the evolution of our universe from its beginning to its end. The quest for a theory of everything is a quest for an understanding of our universe and its origin.

CHAPTER 5

IN THE VERY BEGINNING

In the beginning there was void—a curious form of vacuum—a nothingness containing no space, no time, no matter, no light, no sound.... The curious vacuum held potential. Then the nothingness exploded.... Out of this energy, matter emerged—dense plasma of particles that dissolved into radiation and back to matter. Particles collided and gave birth to new particles. Space and time boiled and formed as black holes formed and dissolved. What a scene![1]

In the above dramatic words, Leon Lederman, Nobel Laureate in physics, described the likely birth scene of our universe. Although there is no sufficient data to show exactly what happened at the time our universe began, the above description is not far from what most cosmologists agree happened.

Quantum Creation from Nothing

Our common understanding of the word *nothing* is the lack of anything, nonexistence, zero, zilch. But when cosmologists talk about the creation of the universe from nothing, it needs further explanation. The best explanation comes from quantum theory, a branch of physics developed in the twentieth century in conjunction with the theory of atoms, particles, and radiation. An important principle of this theory is the Heisenberg's uncertainty principle (discussed earlier in chapter 4). It says that both the speed and the position of a subatomic particle cannot be known precisely at the same time. Measuring accurately one quantity diminishes the precision with which we can determine the other. The same type of linkage has been shown to exist between mass and time. For example, quantum theory predicts that particles can pop in and out of existence from nothing so long as their masses and their times of existence follow the uncertainty principle. In other words, *nothingness* is no longer the lack of anything. Empty space is no longer empty. Even if you remove all the energy from a space, taking out all the matter, all the light, and all the heat, there is still some energy left. The uncertainty principle implies that it is impossible to have an absolutely zero-energy condition. Accordingly, there is always a probability of something (energy or matter) popping out of nothing.

The classical definition of vacuum is a space from which everything that can be removed has been removed. However, in the quantum realm, there is no such thing as a perfect vacuum, a complete void, or an absolute empty space. The quantum theory predicts that the classical vacuum contains fluctuating fields (e.g., electromagnetic field). These fluctuations can give rise to particle-antiparticle pairs[2] if the energy of the fluctuation rises above the rest mass of these particles. The particles appear for very brief times (following the uncertainty principle) and then annihilate each other, thereby reverting back to energy that created them. Because of their sudden appearance and disappearance as part of the energy fluctuation, they are called virtual particles.

The evidence that space is filled with vacuum fluctuations and virtual particle-antiparticle pairs popping in and out of existence out of nothing is seen in a phenomenon called the *Casimir effect*. The phenomenon is named after Hendrik Casimir who predicted in 1948 that two uncharged parallel metallic plates placed in a vacuum would experience an attractive force when brought very close together. The force is measurable only when the separation distance is extremely small, on the order of a few atomic diameters.

The Casimir effect has been measured by several investigators. Its analysis shows that the Casimir effect is caused by radiation pressure from the background electromagnetic field fluctuations (in the so-called nothingness of vacuum) that become unbalanced due to the presence of the plates.[3] As the plates are brought together, the radiation pressure becomes less in the space between the plates than that existing outside the plates. This happens because when the plates are very close, they block out the electromagnetic waves that are too big to fit between the plates. This creates a situation of having more waves bouncing on the outside than from the inside. Hence, the attraction between the plates due to the wave-pressure difference.

Another evidence of quantum fluctuations of the background electromagnetic field is provided by the shift they produce in the frequency of radio waves emitted by hydrogen atoms. The effect is known as the *Lamb-Rutherford shift*.[4] Like the Casmir effect, it demonstrates that there is a background of a fluctuating field of energy in an empty space. Vacuum energy has been linked to a number of other phenomena such as the well-known van der Waals forces (attraction between neutral atoms), the explanation of the observed Planck blackbody radiation spectrum, and the reason for the stability of the hydrogen atom in its ground state from radiative collapse. So, it is a well-established fact that the classical vacuum or empty space is not really an absolute void. It is permeated with fields of pulsating energy governed by laws of quantum mechanics.

The Seed of Universe

It has been shown, based on theory and experimental evidence, that *quantum fluctuations of vacuum energy* can cause the appearance of energetic particles, as allowed by the uncertainty principle. Is it possible that our universe came into being through the same process—from nothing?

In 1973, physicist Edward Tyron[5] proposed that our universe might have originated as a quantum fluctuation of vacuum on a large scale. Just as quantum fluctuations bring virtual particles and antiparticles into existence, this could also have given birth to the seed from which our universe grew. Many cosmologists agree that this is the most plausible route by which our universe came into being.

One objection that is commonly raised to the idea of spontaneous creation comes from our intuitive sense of cause and effect. What caused the quantum fluctuation that sprung up the universe? The answer is that such things happen frequently in the quantum world—the world of atoms and molecules. The commonsensical rules of cause and effect are routinely suspended by the quantum uncertainty principle at the atomic level.

A typical example of a quantum process is the decay of a radioactive material. We can precisely predict the fraction of nuclei that will decay at any time but not which particular nucleus will undergo the transformation. A nucleus may decay immediately, an hour from now, or near the end of the material's radioactive life. The disintegration of a particular nucleus is a random event and cannot be predicted. It should be understood that such a quantum uncertainty is inherent in nature itself. It is a reality that does not follow the principle of what we call "cause and effect" in our macroscopic world. In the microscopic world of almost zero dimensions, similar to what existed at the beginning of the universe, the causation principles were violated in favor of the quantum laws characterized by uncertainty in time, position, and energy.

Let us examine Tyron's idea of spontaneous creation of the universe from the energy conservation point of view. Considering mass and energy to be equivalent ($E = mc^2$), the astronomical data show that the positive energy associated with the total mass of the universe is equal in magnitude to the negative energy locked away in its gravitational field. So, the net energy of the universe is zero. The same applies to the quantum creation of the universe from vacuum. The energy of the quantum particle is balanced by its gravitational field. Thus, no net energy has been created or destroyed.

Tyron's hypothesis that the universe was born from empty space has been questioned by some who ask: where did the empty space come from? In 1982, Alexander Vilenkin extended Tyron's idea of empty space to nothingness. He suggested that the universe started by quantum fluctuation from nothing, defined by the absence of not only matter but also of space and time. Once created from the totally empty geometry, the universe made a transition to a nonempty state of subatomic size through a process called quantum tunneling. That marked the beginning of space, time, and inflationary expansion as modeled in the big bang theory, which will be discussed in the next chapter.

On February 24, 2006, Stephen Hawking delivered the third Dennis Sciama Memorial Lecture on "The Origin of the Universe."[6] He explained that laws of science governed the beginning of the universe. The universe came into existence by spontaneous quantum creation. He visualized the process as bubbles of mini-universes appearing, bursting, expanding, and collapsing. One (or more) survived to grow to a certain size that was safe from recollapsing and continued to expand at an inflationary rate.

There are other scientific theories of how our universe came into existence. But they are variations of the one basic theme, namely, that the universe was created out of nothing. This prompted Alan Guth, the father of the inflation theory, to comment: "It's often said there is no such thing as a free lunch. But the universe itself may be a free lunch."[7]

CHAPTER 6
THE BIG BANG

In the previous chapter, we discussed the theory that the universe began with a quantum fluctuation from nothing. The quantum particle of energy thus created had almost a zero size (like a dot) but was packed with all the energy and matter of the universe to be. In other words, the prenatal universe of such an infinitesimally small size contained all the energy and matter of the presently observed universe, including its one hundred billion galaxies and one hundred billion stars per galaxy. This scenario is mind-boggling indeed.

Because of its almost infinite density, the gravitational field and the space-time curvature of the universe at the beginning must have been close to infinity. Such an enormous gravity could have crushed the baby universe out of its existence were it not for what is known as the *cosmological inflation*. The universe underwent an extremely rapid spurt of growth as soon as it was born. It expanded exponentially and blasted itself out of the quantum realm before it could recollapse. The following sequence of events has been postulated based on a vast amount of astronomical data, analysis of physical forces, and mathematical extrapolation of the expanding universe back in time.

Time Line

At time zero: According to the most recent calculations, our universe was born about 13.7 billion years ago. The starting event is known as the *big bang*. The instant of the big bang has been arrived at by extrapolating the expansion of the universe back in time to its zero size. The space-time started at the big bang. This is the moment when: space-time = 0 (infinite curvature), size = 0, energy density = infinity, gravity = infinity, and temperature = infinity.

At 10^{-43} seconds: The next cosmological landmark is identified when the elapsed time after the big bang reaches 10^{-43} seconds. This is called the *Planck era* (named after one of the most prominent quantum physicists, Max Planck). The size of the universe at this point is a little larger, although still close to zero, but the temperature has dropped from almost infinity to about 10^{32} degrees. Considering the fact that the core temperature of our sun is about twenty-seven million degrees, the temperature of the point-like universe in the Planck era was 10^{24} times that of the sun.

Another important feature of the Planck era was that the four fundamental forces of nature, namely, electromagnetic, strong nuclear, weak nuclear, and gravity (explained in chapter 4) existed as one force, called the *unified superforce*. The entire universe was dominated by quantum uncertainty at this time.

At the end of the Planck era, gravity dissociated itself from the other three forces and space and time became well defined.

At 10^{-35} seconds: At 10^{-35} seconds after the big bang, the universe entered its *inflation phase,* which lasted till time equaled 10^{-32} seconds. This inflation period is characterized by an extremely rapid expansion of space. The inflationary model shows that the universe expanded much faster than the speed of light, almost doubling its size every 10^{-34} seconds.

The inflationary growth spurt was critical for the survival of the universe. The inflation caused the universe to expand too fast for gravity to crush the embryonic universe.

By the end of the inflationary period, at about 10^{-32} seconds from the big bang, the expansion of the universe slowed down to the speed of light. During this time, the universe had expanded from a size of almost zero dimensions to about ten centimeters (from a size smaller than a proton to about the size of a grapefruit).

Another noteworthy thing happened during the inflation era. The strong nuclear force split from its other two partners—the electromagnetic and the weak nuclear—which remained still tied together as the so-called electroweak force.

After the inflationary period, the temperature of the universe dropped down to 10^{27} degrees. The universe at that instant may be imagined as an infinitely hot ball of energy, or strings, expanding at the speed of light in all directions like a balloon.

At 10^{-32} seconds: This moment of time marks the end of inflation and the start of the epoch of electroweak and strong nuclear interactions. The temperature of the universe is about 10^{27} degrees. The energy of particle collisions is about 10^{14} giga electron volts (GeV). The quantum processes cause the radiation to decay spontaneously into particles and antiparticles of matter. Most of the particles and their antiparticles annihilate each other, turning back into electromagnetic radiation. But the laws of physics and the uncertainty principle allow a slight imbalance. At the end of this process, there is a small excess of matter over antimatter.

The critical temperature for the Higgs field to appear is thought to be 10^{15} degrees. So near the end of this epoch, when the universe cools down to this temperature, the Higgs field emerges. It has pervaded the universe ever since.

The concept of Higgs field was discussed in chapter 4. Its role in the context of the big bang theory is to explain how energy and matter got differentiated in the universe at its earliest stages of formation. Particles acquired their mass through interacting with the Higgs field. The stronger the interaction, the heavier the particle, and the weaker the interaction, the lighter is the particle.

Photons do not interact with Higgs field and therefore can travel through it with the maximum possible speed—the speed of light. Material particles, on the other hand, interact with Higgs field, acquire mass, and consequently travel with a speed less than that of light.

The importance of Higgs field lies in the fact that the masses of all particles are based on how much they interact with the Higgs field. In a layperson's terms, one can think of Higgs field as a kind of cosmic molasses that fills all space and causes a type of "drag" on particles that shows itself as mass of the particles.

Another analogy is that of phase changes observed when steam condenses to become water and water freezes to become ice. Matter is like "frozen" energy. The Higgs field, which can only exist at temperatures of 10^{15} degrees or less, acted as a catalyst in the creation of material particles from energy after the inflationary period when the temperature of the universe dropped down to 10^{15} degrees or less.

The Higgs field continues to play its part in giving particles their mass property. All space is filled with it, and yet it is indistinguishable from empty space. No wonder the carrier of this field is called "The God Particle."[1]

At 10^{-10} seconds: A small excess of matter over antimatter that resulted after the inflationary period and the emergence of Higgs field gave rise to the fundamental particles such as quarks and electrons. At about 10^{-10} seconds after the big bang, the electroweak force split into the electromagnetic and the weak nuclear. This epoch is characterized by the separation of all forces of nature—gravity, electromagnetic, strong nuclear, and weak nuclear.

The temperature of the universe at 10^{-10} seconds after the big bang is about 10^{15} degrees and dropping rapidly as the universe is expanding at the rate close to the speed of light. The size of the universe by now has increased enormously—about 10^{15} centimeters in diameter, which is comparable to the size of our solar system.

At 10^{-4} seconds: The fundamental particles, quarks, begin to assemble and form particles such as protons and neutrons, the building blocks of atomic nuclei. So this time marks the beginning of the formation of preatomic particles.

At one hundred seconds: By this time, the universe has expanded to many times the size of our solar system and the temperature has dropped enough to allow protons and neutrons to come close together to form nuclei. The strong nuclear forces, which are very powerful but short-ranged, play their part in "gluing" protons and neutrons together without getting stripped apart by energetic radiation.

It is amazing that just after one hundred seconds following the big bang, the material fabric of the universe, because of its rapid expansion and cooling, allowed the synthesis of light atomic nuclei. Hydrogen was the first element to form because its nucleus consists simply of a proton. The next element to form in a significant proportion was helium, which has a nucleus containing two protons and two neutrons. A small amount of lithium, which has a nucleus containing three protons and four neutrons, was also formed around that time. Although at this point in time, the universe was cool enough to form nuclei of light elements, it was still too hot to allow the formation of atoms (nuclei with electrons revolving around them).

At five hundred thousand years: It took about five hundred thousand years for the universe to cool down enough to let the nuclei capture electrons. Soon all the electrons got bound up in atoms. This created conditions of radiation to stream freely through interatomic spaces without being constantly scattered by electrons. Scientists have measured this radiation that now appears as *cosmic microwave background.*

The cosmic microwave background radiation (CMBR) is considered as fossil radiation because it was generated about five hundred thousand years after the big bang and it has remained intact even up to this day. The temperature of the universe when it was five hundred thousand years old was about six thousand degrees Celsius (comparable to the temperature of sun's surface). The heat radiation emitted at that temperature is expected to have a wavelength of about 10^{-6} millimeters, which is the range of ultraviolet light. But the expansion of the universe for 13½ billion years since has stretched the wavelength of that radiation to about one millimeter (in the microwave range). From the physics of what is known as "blackbody radiation,"[2] it can be shown that heat radiation of this wavelength corresponds to a temperature of about -270 degrees Celsius (close to absolute zero, which corresponds to 0 degree Kelvin or -273 degrees Celsius). That means the average temperature of the universe has dropped from 6000 degrees Celsius when it was five hundred thousand years old to about -270 degrees Celsius at the present time. The fact that these calculations agree with the average temperature of our current universe provides a strong support to the big bang model of the universe.

At one billion years: After the era of atomic formation and their congregation, it took about a billion years for matter to condense into stars and galaxies. Gravity had a major role to play in their formation and configuration. Indeed until then, the universe was a dark place with no light that could be visible with the human eye.

Before the formation of stars and galaxies, the matter in the universe consisted of approximately 75 percent hydrogen and 25 percent helium. The heavier elements that we observe today, including those that constitute Earth and life on Earth, were created later in the interiors of the stars through the process of fusion and spread throughout the universe by supernova explosions. However, in the cosmic scheme of things, they may be considered as only trace impurities.

At five billion years: This era marks the time when our solar system was formed. From the expansion of the universe after the big bang, the formation of stars and galaxies, and the analysis of their elemental composition, it has been estimated that our solar system (the sun with its planets, including the earth, and their moons) came into being about five billion years ago. The size of the universe at this time was approximately two-thirds of its present size.

At 13.7 billion years—the present: Formation of stars and galaxies continues to this day, and the old ones are burning up or receding from each other like figures on the surface of an expanding balloon. It is a dynamic universe that is governed by physical laws, and we have plenty of time (billions of years, if we survive that long) to study it and make predictions about its future.

Evidence

The big bang (also called the hot big bang) model rests on three main pieces of scientific evidence:

1. Observed expansion of the universe

2. Detection of cosmic microwave background radiation (CMBR) and its correlation with the average temperature of the universe

3. Relative abundance of hydrogen and helium

Because the current astronomical observations and data have to be extrapolated back in time to establish singularity, an ironclad proof of the big bang theory is not possible. However, by the application of established mathematical and physical principles, a model has been constructed that predicts our current observations and measurements. Like any scientific theory, it is not irrefutable, so the scientific investigation is expected to continue in order to get as close as humanly possible to the truth of creation.

Cosmic expansion: Ordinary white light consists of a spectrum of colors. We can observe these colors by passing a beam of sunlight through a glass prism or by looking at the rainbow that occasionally forms in the horizon opposite the sun after a rainfall. The light is split into its constituent colors after passing through the prism or the raindrops.

White sunlight is a combination of colors: violet, indigo, blue, green, yellow, orange, and red. Each color is characterized by the wavelength or frequency of its photon. In going from violet to red, the wavelength increases, or the frequency decreases. If you take a picture of the white-light spectrum from a stationary source of light, the colors appear at specific locations, characteristic of their wavelength. However, if the object is moving relative to the observer, the wavelength (or frequency) of each color changes. As a result, the colors of the spectrum get shifted toward the blue or the red end, depending on whether the object is approaching the observer or receding from the observer. This phenomenon is known as the Doppler effect, which we experience in our ordinary lives when an ambulance passes us by. The pitch, which is related to frequency of sound, increases when the ambulance is approaching, and it decreases after the ambulance passes. The magnitude of the change in pitch depends on the speed of the ambulance relative to the observer. In other words, the Doppler shift gives us an indication of relative speed. A traffic policeman similarly uses radar (microwave beam) to determine the relative speed of a car by the same principle. In fact, the Doppler shift in the wavelength of electromagnetic radiation is one of the most useful and accurate methods of measuring the relative speed of objects.

In 1929, Edwin Hubble showed that the light from distant galaxies is shifted toward the red end of the spectrum. That means these galaxies are moving away from us. The earth does not occupy a central position in the universe. We are like a dot on a spherical balloon. As the balloon expands, all dots move away from each other irrespective of their position. However, it should be mentioned that as the universe is expanding, the distance between the galaxies is increasing but not the distance between the objects within the galaxies themselves. The stars, planets, gaseous clouds, or black holes in a galaxy are held together by mutual gravity, and they are moving as a group relative to the neighboring galaxies.

Hubble related the redshift in the light from a galaxy to its speed of recession. He also measured the distance of galaxies by using a method called the Cepheid method. Cepheids are stars whose brightness varies regularly over a period of days or weeks. Earlier in 1912, astronomer Henrietta Leavitt had shown that the period of a Cepheid's variability is related to its intrinsic brightness. By comparing the intrinsic brightness of a Cepheid with its apparent brightness, one could calculate its distance by measuring the degree of dimness caused by distance.

By comparing the speed of recession with the distances to the galaxies, Hubble discovered that the farther away the galaxy was, the faster it receded. From the redshift data (related to the speed of recession) of galaxies and their distances, he formulated a relationship, now known as *Hubble's law*. This law states that the speed of recession of a galaxy is simply given by its distance multiplied by a constant number called the Hubble's constant. Mathematically,

$$v = Hd$$

where v is the velocity of recession in kilometers/second, d is the distance in million light years, and H is the Hubble's constant in kilometers/second/million light years. The most currently measured value of H is about 18.5 kilometers per second per million light years of distance.

Hubble's discovery means that the universe (space-time) is expanding like a loaf of raisin bread when it is baked. As the dough rises, the distance between raisins increases. Raisins in this case are analogous to the galaxies or galaxy clusters. As stated earlier, the expansion of the universe is increasing the intergalactic space but not expanding the galaxies themselves. Their stars are held together in a group by gravity. The chemical and gravitational forces that bind planets, stars, galaxies, or galaxy clusters together are stronger than the forces of cosmic expansion.

It has been observed that the redshift of a distant galaxy is greater than that of one closer to Earth. That means more distant galaxies are receding at a greater speed, as predicted by Hubble's law. This is, of course, true for any point in the universe where such observations are made, since the earth is neither the center nor a unique location in the universe. The universe is like the surface of a balloon. No point on its surface constitutes the center of expansion, nor is there any edge to the universe.

The redshift data indicate that some of the very distant galaxy clusters and quasars are moving away from us at greater than 90 percent of the speed of light.

There seems to be no sign of this expansion slowing down in the foreseeable future.

In 1915, Einstein presented his general theory of relativity in which he described gravity as a distortion of the space-time by massive objects. There was yet another important aspect of his theory that related to the dynamics of the universe. The equations of general relativity showed that the universe could not be static. It could either shrink or expand, but it could not be static. As Einstein struggled to reconcile his equations with the prevailing view of a static universe, he finally modified his equations in 1917 to include a new term: the *cosmological constant*. This term represented some kind of a repulsive force that would balance the pull of gravity, thereby making the universe static. But many years later, Hubble and others proved beyond a doubt that the universe was actually expanding at an enormous speed. Einstein called it, "the greatest blunder of my life."[3]

The significance of cosmological constant to the observed cosmic expansion has been investigated beyond its mathematical necessity. Some theorists believe that the cosmological constant could have been very high in the beginning. The implication is that some mysterious "dark energy" provided an immensely powerful antigravity or cosmic repulsion in the beginning and then quickly switched off. Ordinary gravity was overwhelmed initially, causing exponential expansion of the universe. Part of this repulsive dark energy could also have been converted into heat energy of the big bang.

The fact that the universe is expanding makes a strong case for a singularity of the big bang. If you take the current motion picture of receding galaxies and run it back in time, the galaxies would appear to converge on to a single point. The moment of that convergence coincides with the moment of the big bang.

The cosmic microwave background radiation (CMBR): In 1948, George Gamow and his students predicted that if the big bang theory were correct, one would expect to find relics of heat radiation emitted when the universe was roughly five hundred thousand years old with a temperature of about six thousand degrees Celsius. Because of the expansion, the universe has since cooled down to a few degrees above absolute zero and the wavelength of the original heat radiation has accordingly been stretched to the microwave range. Detection of this radiation could therefore be a big evidence for the big bang theory.

It was not until 1965 that CMBR was discovered by Arno Penzas and Robert Wilson of Bell Laboratories who adopted their radio antenna for satellite communications for use in cosmic research. The CMBR appeared as a faint noise that they could not eliminate, even after cleaning their antenna horn of all the pigeon droppings. The detected radiation had a wavelength characteristic of heat radiation emitted at about -270 degrees Celsius, a temperature very close to that of the present universe, on the average. Penzas and Wilson received the Nobel Prize in 1978 for their discovery. It is considered to be one of the most direct evidences for the big bang theory.

Further evidence was provided by NASA's Cosmic Background Explorer (COBE) satellite in 1989. It measured the whole spectrum of the CMBR. The analysis of the data confirmed that the background radiation represented the relics of a perfect blackbody heat radiation spectrum that existed in the universe when it was several thousand degrees hotter than today.

Relative abundance of light elements: As the universe cooled down to low enough temperatures to allow formation of atomic nuclei, the first element to form was the lightest of all—the hydrogen. The nucleus of hydrogen consists of a single proton. Since protons are slightly lighter than neutrons, nature preferred protons to neutrons because the former have a lower energy state. As a result, an excess of protons in the hydrogen nuclei began to accumulate. When the temperature of the universe dropped to about ten billion degrees Celsius, the ratio of free protons to neutrons became fixed at about seven to one.

As the temperature dropped below three billion degrees, neutrons began to pair with protons to form nuclei of heavy hydrogen (also called deuterium) with a nucleus of one proton and one neutron, but the universe still contained mostly hydrogen nuclei—the unpaired protons.

With further cooling of the universe to about one billion degrees Celsius (approximately three minutes after the big bang), the deuterium nuclei started to pair up to form helium nuclei (two protons plus two neutrons). It was not long after the big bang that the material content of the universe stabilized at 75 percent hydrogen and 25 percent helium. The fact that about the same proportion is observed today provides another piece of evidence in support of the big bang model.

CHAPTER 7
COSMIC EVOLUTION

In this chapter I describe the formation, evolution, and fate of our universe with its galaxies of stars, planets, and other familiar heavenly bodies. The intent is to give the reader an appreciation of how the physical laws gave rise to what the scriptures call the heavens and the earth. If science can explain the creation of these heavenly objects, do we need a creator? Before we answer this or other similar questions involving faith and science, we need to understand how the universe originated and evolved to be what it is today.

John Barrow paints the following picture of the cosmic evolution:
In the beginning, the universe was an inferno of radiation too hot for any atoms to survive. In the first few minutes, it cooled enough for the nuclei of the lightest elements to form. Only millions of years later would the cosmos be cool enough for whole atoms to appear, followed soon by simple molecules, and after billions of years by the complex sequence of events that saw the condensation of material into stars and galaxies.[1]

The evolution story of our universe starts at the beginning when all its matter and energy were concentrated into an infinitesimally small point. It was infinitely dense and infinitely hot. Space and time were nonexistent.

Then the universe exploded. Time began to tick and the universe began to expand. The expansion caused the temperature to drop, and some of the energy began to condense into matter.

The early moments of the universe, as described in chapter 6, are characterized by its expansion and cooling at an incredible rate. Right after 10^{-35} seconds of the big bang, the universe was expanding much faster than the speed of light, doubling its size every 10^{-34} seconds. This is known as the inflationary period and was critical for the survival of the baby universe. It expanded too fast for gravity to crush the embryonic universe. There is some evidence that the inflationary expansion could have been facilitated by a mysterious energy, called dark energy, which is repulsive in nature and antigravity. We do not know a whole lot about this strange kind of energy, but it fits the model of our expanding and accelerating universe in spite of the gravitational forces generated by the visible and the dark matter in the universe.

The inflationary period was essentially over by 10^{-32} seconds, but the universe continued to expand and cool at a lower but still exorbitant rate. By about 10^{-10} seconds, it was expanding at close to the speed of light. By now the universe had grown in size comparable to our solar system (about ten billion kilometers in diameter). The temperature had gone down considerably, but it was still close to 10^{15} degrees Kelvin.

As soon as the temperature dropped down to 10^{15} degrees Kelvin, the mysterious Higgs field kicked in. As a result, energy went through a phase change. Particles formed out of energy and acquired their mass through interaction with the Higgs field.

The Higgs field has pervaded the universe ever since it emerged. It mediated the creation of particles out of energy and gave them their mass properties. The fundamental particles, quarks, came into existence and began to form protons and neutrons, the building blocks of atomic nuclei. The epoch of preatomic particle formation started as early as 1/10,000 of a second after the big bang.

As the universe expanded to many times the size of our solar system (at about one hundred seconds), the temperature dropped enough to allow protons and neutrons to form light nuclei—hydrogen and helium. At this time, the hydrogen nuclei were most abundant, making up 75 percent of all the matter in the universe. Helium constituted the other 25 percent. This proportion has remained the same up to this day with only trace amounts of other nuclei.

It took close to three hundred thousand years for the universe to cool down enough to allow nuclei to capture electrons and become atoms. Until then, the radiation was constantly scattered by electrons. But as soon as the electrons got captured into their orbits around the nuclei, radiation could stream freely through interatomic spaces. Figuratively speaking, the electronic fog condensed over the nuclear dust and made the universe more transparent to radiation. The relic radiation glow of that era still pervades the universe and can be measured by antennas tuned to cosmic microwave background radiation.

Birth of Stars

After about a billion years of expanding and cooling, certain regions of the universe contained more atoms than the others. The regions that were denser than the average experienced extra gravitational attraction and slowed down in their expansion. Eventually, gravity stopped their expansion and caused them to collapse. However, by the gravitational pull of matter outside, the collapsing regions developed a rotation to balance the attraction of gravity. This is a simple description of how the rotating galaxies were born.

The gas in the galaxies was uniformly distributed. With time it became clumpy and formed patches of dust clouds that collapsed under their own gravity. As the gravitational pressure increased in a cloud, the temperature in it reached millions of degrees (the center of the sun is about sixteen million degrees). At such high temperatures the fusion reactions started. The hot compressed clouds began to convert hydrogen into helium. As a result, enormous amounts of energy were released in the form of heat, light, high-energy gamma rays, neutrinos, and other particles. The outward pressure of radiation thus generated balanced the inward pressure of gravity, thereby forming stable stars within the galaxies.

It is estimated that the universe contains about one hundred billion galaxies and each galaxy contains about one hundred billion stars. The most distant object in the universe that has been recently discovered is a small galaxy that lies approximately thirteen billion light years from Earth (one light year = 9.5×10^{12} kilometers = 5.7 trillion miles). Since the current age of the universe is about 13.7 billion years, this means the light reaching us now originated from that galaxy when the universe was only one hundred million years old. It is a much smaller galaxy, just two thousand light years across, compared to our Milky Way, which is roughly one hundred thousand light years in diameter.

A galaxy contains stars, gas, and dust—all held together by gravity. The star formation occurs in the dust clouds, as described earlier. The clouds of interstellar gas and dust are called *nebulae*. Some of them emit light of their own, while others reflect light or appear dark against the lighted background of more distant stars and nebulae.

The astronomical data suggest that the star formation was a gradual process that reached its peak when the universe was about seven to nine billion years old. The galaxies were then hyperactive and gave rise to bright star-like objects, called *quasars*.

The quasi-stellar objects, or quasars, are bright energetic nuclei of distant galaxies. These galaxies are so far away that their brilliant central regions, the quasars, look like bright stars—much brighter than the entire surrounding galaxy. The mechanism of their formation is thought to be the gravitational collapse of not just one star but of the whole central region of a galaxy.

Death of Stars

All stars are formed through the same evolutionary process, but their demise follows different routes depending on their size. A star can live for billions of years until it runs out of fuel. Then it meets a dramatic stellar death. The more massive the star, the more dramatic is its demise.

As discussed earlier, a star generates energy by burning (converting) hydrogen in its core to helium by the process of nuclear fusion. Near the end of its life, when the hydrogen in its core is depleted, it starts burning hydrogen, by nuclear fusion, in its envelope or shell. As the hydrogen burning progresses gradually outward, the star's outer layers expand and begin to cool off. For example, a relatively small-mass star, like our sun, will turn into a *red giant* when it is finished with its core hydrogen burning. Its temperature will cool down to 3,500 degrees Kelvin and it will swell out to a gigantic size, gobbling up its nearby planets, Mercury, Venus, and Earth.

After burning all its hydrogen in the core and becoming a red giant, a small-mass star like the sun will collapse under its own gravity and compress the matter to a degenerate state in which electrons and nuclei are packed as tightly together as possible by the star's gravity. When a star reaches that stage, it is called a *white dwarf.* A white dwarf eventually fades from view as it cools.

A more massive star (e.g., a star more than ten times heavier than the sun), on the other hand, becomes a *supergiant* when it exhausts its hydrogen fuel in the core. Then it collapses and compresses the core to a high enough pressure and temperature to start the nuclear burning of helium. A massive supergiant can go further up the nuclear fusion chain with the burning of lighter elements into heavier ones like beryllium, carbon, and oxygen until it creates iron, which is the heaviest metal produced in the stars through this process.

After burning all its fuel, a massive star collapses under its immense gravitational pressure so as to overcome the electron degeneracy pressure—the pressing of electrons against the nucleus. It will continue collapsing until neutron degeneracy pressure will stop further collapse. At this stage, it is called a *neutron star* and consists almost entirely of neutrons. Neutron stars typically have a diameter of about twenty kilometers and densities of about 10^{17} kilograms per cubic meter.

Rapidly rotating neutron stars with intense magnetic fields are known as *pulsars.* They are characterized by strong radio pulses emitted with a repetition frequency of 0.001 to 4 seconds. If correctly oriented, their narrow beams of radio emissions sweep the earth like a lighthouse beacon.

Another fate of a massive star is a catastrophic stellar explosion after it has exhausted its fuel. As the star collapses to a denser and denser state, the core can become unstable and collapse suddenly in less than a second. When the matter reaches nuclear densities, it cannot take further pressure. However, because of the suddenness of the implosion and resistance to further collapse, the core bounces back and a shock wave is generated outward. The outer layers of the star are blown off in a spectacular explosion, leaving behind a neutron star. Such an exploding star is called a *supernova*. These explosions are so huge that the supernova can outshine an entire galaxy.

By the time a supernova is ready to blow off its outer layers, the star has developed an onionskin structure with layers containing various products of its nuclear alchemy such as carbon, oxygen, silicon, and iron. Heavier elements in the periodic table are produced only in trace amounts; some in the heat of the supernova explosion itself while others through nuclear transformation by capturing neutrons in the red giant stars. Thus, the supernovae not only manufacture various elements when they are stars, they disperse them in the universe when they explode.

Some stars are so massive that even the neutron degeneracy pressure cannot prevent their collapse. Their force of gravity is so strong that it overcomes any opposing resistance to collapse. The star then continues to collapse on itself until it forms a *black hole*. The black holes are so dense and can generate so much gravity that even light cannot escape—thus the name black hole.

There is a lot of interest in the physics of black holes. In their work on black holes between 1965 and 1970, Penrose and Hawking showed that there is a singularity of infinite density in black holes like the big bang at the beginning of time. Black holes are characterized by mass, electric charge, and angular momentum. The gravity is so strong in a black hole that any matter or radiation that comes too close to the black hole is instantly sucked in to a physical singularity at its center.

Our Solar System

The term solar system applies to the sun and objects gravitationally bound to it. Besides the sun, which is the centerpiece of the system, the group includes planets and their satellites, comets, meteors, asteroids, and interplanetary dust and gas.

The current view is that the sun and the other objects in our solar system came into being at the same time, about five billion years ago when the universe was two-thirds of its present size. They were formed from solar nebulae, a cloud of interstellar dust and gas. The birth of the sun from this nebula followed the same process as that involved in the birth of any star, namely, compression of the center due to gravity and increase in the pressure and temperature of the core so high that the nuclear fusion of hydrogen into helium would occur.

As the center of the solar nebulae compressed and gave birth to the sun, the surrounding dust and gas stayed cooler the farther away they were from the sun. The elements such as metal, rock, and ice began to condense out of the cloud at different rates, depending on their distribution and distance from the protosun. The larger condensed masses vacuumed up the smaller ones through their greater gravitational attraction, thus giving rise to planets of different masses and densities.

All the planets of the solar system rotate around the sun in a counterclockwise direction. In addition, the sun, the planets, and the majority of their moons spin on their own axes in the same direction. The fact that the planets orbit the sun in almost the same plane suggests that they were all formed from the same rotating disk of the solar nebula.

Our sun has eight planets. In the order of their closeness to the sun, these are Mercury, Venus, Earth, Mars, Jupiter, Saturn, Uranus, and Neptune. Pluto is no longer considered a planet by consensus of a body of scientists. It is classified as a dwarf planet.[2] In addition, there are rocky bodies called asteroids that also orbit the sun. Most of them lie in the asteroid belt between Mars and Jupiter.

Some planets have satellites or moons of their own while others don't have any. The earth has one satellite, the moon. The planets also differ in their composition, size, density, temperature, and atmosphere. We will briefly describe them here to familiarize the reader somewhat with these fascinating objects—our closest neighbors in the sky. They are also called heavenly bodies and have been a major part in folklore and religious mythology in all human civilizations.

The sun: The sun is our nearest star. It is approximately 149 million kilometers (93 million miles) from the earth. The light from the sun takes eight minutes and nineteen seconds to reach the earth. The diameter of the sun is about 1.4 million kilometers (870 thousand miles).

Like most stars, the sun consists mostly of hydrogen. Its composition is approximately 74 percent hydrogen, 25 percent helium and 1 percent other elements. Its mass is about 2×10^{30} kilograms. Although the sun is huge compared to the earth, among stars it is considered to be of medium size.

The core of the sun, where nuclear fusion is taking place, is at temperatures on the order of fifteen million degrees. Enormous amounts of energy in the form of heat and radiation are being released as it converts about six hundred billion kilograms of hydrogen into helium per second.

The sun is not a solid body. It is what physicists call plasma—an ionized state of gas consisting of free electrons, protons, and charged ions. Neutral atoms are absent because of the intense heat.

As the sun rotates about its own axis, the rotating body of its plasma generates a magnetic field of roughly the same strength as the earth's. However, the turbulence in the top layers creates pockets of intense magnetic fields that are responsible for some of the phenomena observed in the sun's activity.

Mercury: Mercury is named after the messenger of gods—god of commerce, cleverness, and thievery—identified with the Greek god Hermes. Because of its relatively close proximity to the sun, Mercury is the fastest orbiting planet in the solar system. It takes just eighty-eight earth days to orbit the sun. It also rotates around its own axis, completing one rotation in about fifty-nine days.

Mercury is much smaller than Earth. Its diameter is about 4,900 kilometers (3,000 miles). Most of its surface is heavily cratered, presumably by impacts of meteorites, asteroids, and comets.

Mercury is mostly made up of iron. About three-quarters of its inner diameter consists of an iron core surrounded by a mantle of rock and thick crust. It has practically no atmosphere, which is almost a vacuum with only trace atoms of hydrogen, helium, oxygen, sodium, and potassium.

Being the closest planet to the sun, Mercury suffers from the widest extremes of temperature. Its temperature can soar up to 470 degrees Celsius at noon and dip down to as low as -173 degrees Celsius at night.

Venus: Venus is named after the goddess of love and beauty—identified with the Greek goddess Aphrodite. A view from Earth shows Venus to be the most brilliant planet in the solar system. It prominently appears in the morning as the "morning star" and in the evening as the "evening star." In spite of its being named after the goddess of love and beauty, it is nothing but a burning, suffocating place that you could literally call hell. It is completely devoid of water, with surface temperatures reaching as high as 480 degrees Celsius. The atmosphere consists of Venusian clouds composed of carbon dioxide interspersed with sulfuric acid droplets. The gas is dense and exerts a pressure ninety times that of Earth's atmosphere at the surface.

The diameter of Venus is about 12,000 kilometers (7,500 miles). Its composition is similar to that of Earth, namely, iron and nickel in the core surrounded by silicate rocks and other elements. The images of Venus's surface show crisscrossing of faults and fractures, tall mountains, and lowland plains of volcanic origin. It is not known if any of the volcanoes are active today.

Earth: Our planet is about 4.6 billion years old. It is the only planet known in the universe to sustain life. Its composition, temperature, atmosphere, and location in relation to the sun all seem to strike a perfect balance to sustain life.

Earth's vital statistics are approximately: diameter, 13,000 kilometers (8,000 miles); circumference, 40,000 kilometers (25,000 miles); average distance from the sun, 149 million kilometers (93 million miles); time to orbit the sun, one year; orbital speed, 2,000 kilometers per hour (1,200 miles per hour); time of rotation around its own axis, twenty-four hours; rotation speed, 1,700 kilometers per hour (1,000 miles per hour); and tilt of its rotation axis, twenty-three degrees relative to its orbital plane.

While the rotation of the earth gives rise to day and night, the tilt of its spin axis is responsible for the four seasons. For the part of the year that either of the two hemispheres is tilted toward the sun, it receives more sunlight and has longer days and shorter nights. That means there is summer in that hemisphere while the other hemisphere experiences winter. The seasons reverse when Earth's axis tilts away from the sun.

Some of the most vital and unique features of Earth are its oceans of water that cover about three-quarters of its surface area. Geological data show that oceans have existed for nearly four billion years. It is thought that the oceans were formed by the condensing of steam as Earth cooled and the melting of ice in the comets that impacted our planet.

Earth's atmosphere is critical in supporting the planet's many life-forms. Its average composition is 78 percent nitrogen, 21 percent oxygen, 0.9 percent argon, and 0.03 percent carbon dioxide. The atmosphere also contains water vapor, continuously varying in amount as the oceans evaporate, clouds form, and rains fall in a cyclic fashion.

The normal atmospheric pressure at sea level is 101.3 kilopascals (1 pascal is 1 newton of force exerted per square meter of surface area). Since the atmospheric pressure causes the mercury barometer to rise, the normal pressure at sea level is also expressed as 760 millimeters (29.9 inches) of mercury. The atmospheric pressure drops with an increase in altitude above sea level at an approximate rate of one inch of mercury per one thousand feet.

As we all know, Earth's atmosphere is not static. Like oceans, it is in a state of constant circulation, driven largely by temperature differences in various regions of the earth. The temperature difference creates a pressure gradient that causes air to move to equalize the pressure. Strong winds can turn into hurricanes or typhoons, which churn the ocean water to aerate it and provide the needed oxygen to sea life. The consumption of oxygen by earthen life is likewise balanced by the production of oxygen by trees and vegetation that convert carbon dioxide into oxygen. There is indeed an ecosystem that exists between life and the physical and chemical environment in which it is sustained.

From the observation of interrelationship between life and environment, it seems as if nature has deliberately created this ecosystem for life to exist and thrive on Earth. Is it a part of a deliberate design by a supernatural power (e.g., God) or does it simply exist as a result of billions of years of evolution? We are not ready to address that question at this point. It will be addressed later in the book.

The core of the earth, which contains about one-third of Earth's mass, is mostly iron with a small percentage of nickel, sulfur, oxygen, and some other light elements. The core temperature, which may reach as high as five thousand degrees, is maintained by radiation generated by its various radioactive elements. The crust of the earth contains mostly silicon, iron, magnesium, calcium, aluminum, sodium, potassium, and oxygen. There are ninety-two different elements that are known to occur naturally in the earth.

The moon: As a child I was fascinated by the moon—it looked like a happy human face staring at me with a smile. When I walked, it walked with me. When I ran, it ran with me. I really thought that the moon followed me everywhere because it liked me. I had a celestial friend!

As a grown-up, I still consider the moon my friend—a serene and peaceful face with an everlasting smile. But every now and then, I begin to think about what it really is—an object like the earth but without oceans, forests, or life. Its mystery disappears when I visualize Neil Armstrong bouncing on its powdery surface and hoisting the American flag. Then I picture it as a remote island in space, fascinating and exotic but devoid of any life—a cold, lifeless face with a smile!

Many planets that orbit the sun have moons that revolve around them. Earth has one moon—our moon. The moon also rotates on its own axis in approximately the same time as it orbits the earth. That is why it always shows the same hemisphere or face to the earth.

The moon is a rocky, spherical body with no light of its own. It shines only by reflected sunlight. We see its whole face (hemisphere) when it is opposite the sun. But when it is in the same direction as the sun, its hemisphere facing us is unlit and therefore invisible. In between these positions, we see only a part of the hemisphere illuminated.

Analysis of the data collected from lunar samples suggests that the moon was originally a part of the earth. It was formed by a giant impact of a large object about 4.5 billion years ago. The rocky debris that was thrown out into space by the impact coalesced together by gravity to form the moon.

The moon has a diameter of about 3,500 kilometers (2,100 miles). It rotates around its own axis in 1.0 earth day and around the earth in 27.3 earth days. Its surface gravity is only 17 percent of that found on the earth. The average daytime temperature is -130 degrees Celsius.

The dark patches on the moon's surface are known to be low-lying areas that were formed 3–4 billion years ago by volcanic lava flow. The other notable features, the craters, were caused by the impact of comets and asteroids on the moon's surface. These craters have remained preserved because of the lack of atmosphere around the moon. There is nothing to wear away the old craters.

Mars: Mars is named after the god of war—identified with the Greek god Ares. Mars is the closest planet to Earth. Its average distance from the sun is about 228 million kilometers (141.6 million miles). Because of its relative proximity to Earth, it is readily observable in the sky with even a small telescope.

Mars is notable for its red color, which it gets because of the iron-rich composition of its rocks and dust. It is approximately 6,800 kilometers (4,200 miles) in diameter, which is about half the size of Earth. Its diurnal rotation is about twenty-four hours, thirty-seven minutes, and its orbital rotation around the sun is about 687 earth days.

Both polar caps of Mars are covered with ice. Mars experiences seasonal dust storms that are caused by the temperature difference between the polar caps and the relatively warmer equatorial regions closer to the sun. These dust storms can gust up to 90 kilometers (56 miles) per hour.

Data from *Mariner 4, 6,* and *7* in the 1960s have shown Mars to have a thin atmosphere of carbon dioxide with a pressure of less than 1/100 of that of Earth. The average annual temperature ranges from -58 degrees Celsius at the equator to -123 degrees Celsius at the poles. At such low temperatures and low pressures, water cannot exist in a liquid form on the Martian surface. However, the *Mariner* and *Viking* pictures have revealed evidence of water flow on the surface presumably in its early history.

The Martian surface is heavily cratered in the southern regions and contains low-lying plains in the northern hemisphere. There is some evidence of ancient action of water and lava flow in the northern plains.

Although there is no liquid water currently on the surface of Mars, there is a great deal of frozen water locked up in the polar caps and deeper beneath the surface. From the *Mariner* and *Viking* pictures, it is speculated that Mars must have had a denser and warmer atmosphere with flowing water on its surface a few billion years ago.

This brings us to the question whether life ever existed on Mars. Current thinking is that there is insufficient evidence to answer this question. However, if the planet were warmer and wetter in its early history, there is a probability that life could very well have existed there.

Jupiter: Jupiter is named after the supreme god—identified with the Greek god Zeus. Jupiter is the largest planet of the solar system. Its mass is greater than twice that of all other planets combined. Its distance from the sun is about 780 million kilometers (483 million miles). Its diameter is about 143,000 kilometers (89,000 miles), which is about eleven times that of Earth.

Jupiter's period of revolution around the sun is about 11.86 earth years. Its period of rotation around its own axis is a little less than ten hours. Because of its high speed of spin, Jupiter has an appearance of swirling clouds with waves and giant eddies, characteristic of a dynamic weather system.

Jupiter is a gaseous planet but has a solid core of silicate ice, surrounded by liquid metallic and molecular hydrogen. The atmosphere is made up of about 86 percent hydrogen and 14 percent helium, with trace amounts of ammonia, water vapor, and other compounds. Various layers of the atmosphere have different temperatures and pressures. The coldest level is -153 degrees Celsius at the top of the clouds, and the hottest level is at the surface of the planet, which is covered by liquid metallic hydrogen, at 10,000 degrees Celsius. Metallic liquid is a substance in which atoms are stripped of their electrons and form a conducting liquid metal.

The famous Great Red Spot of Jupiter is a large atmospheric eddy, a cloudy vortex rotating in counterclockwise direction. It is variable in size but can measure as much as 14,000 kilometers by 40,000 kilometers (8,700 x 25,000 miles), which is larger than Earth's diameter.

Jupiter is known to have sixteen moons, which have been named as Io, Europa, Thebe, Callisto, Ganymeda, Metis, Amalthea, Adrastea, Ananke, Carme, Sinope, Pasiphae, Leda, Lysittea, Himalia, and Elara. In addition to the moons, Jupiter is surrounded by a system of thin rings consisting of small rocky particles.

Saturn: Saturn is named after the god of agriculture—identified with the Greek god Cronus. Saturn is the second largest planet in the solar system. It is best known because of its spectacular ring system. It orbits around the sun in about 29.4 years and rotates in about ten hours. Its diameter is about 120,000 kilometers (75,000 miles).

The core of Saturn consists of silicate and rock material surrounded by liquid metallic hydrogen and liquid molecular hydrogen. The surface gravity is close to that of Earth, and the average surface temperature is about -180 degrees Celsius.

Saturn's atmosphere is similar to that of Jupiter, being 80–90 percent hydrogen and 10–20 percent helium with traces of other gases including methane and ammonia. Near the equator, winds blow at very high speeds at the top of the clouds—about 2,000 kilometers per hour (1,200 miles per hour).

The ring system of Saturn is made up of seven main rings, A–G. The largest of these is the E ring, with a width of about 300,000 kilometers (186,000 miles). The rings are made up of many thin ringlets and contain small particles ranging in size from a centimeter to several meters. The particles are composed of rock and water ice.

Saturn has eighteen satellites or moons. Of these, Titan is the largest. Titan, in fact, is the second largest moon in the solar system after Jupiter's Ganymede and the only one with a thick atmosphere. Its atmosphere is 90 percent nitrogen and 10 percent other gases such as methane.

Uranus: Uranus is named after a Greek god—the husband or son of Gaea (Earth) and father of the Titans, Furies, and Cyclops. He was overthrown by his son Cronus (Saturn). Uranus is an aqua-colored planet with a diameter about four times that of Earth. It is tilted on its axis through about ninety-eight degrees. As a result, the planet moves around the sun on its side with its rotation axis lying almost in the plane of its orbit. This causes the planet to suffer the most extreme seasonal changes of all the planets in the solar system.

Uranus is a gaseous planet like Jupiter and Saturn, covered with thick layers of clouds consisting mainly of hydrogen and helium. The pressure rises with depth and turns hydrogen and helium from a gaseous state to a liquid state. With further increases in depth, the liquid is compressed into a solidified layer. The center of Uranus is highly compressed into a core of rocky material.

The atmosphere of Uranus is similar to the other giant gaseous planets, consisting mainly of hydrogen (80 percent), some helium (15 percent), and other gases such as methane and hydrocarbons. Absorption of red light by methane causes the reflected sunlight from the planet to appear as greenish blue.

The average surface temperature of Uranus is about -216 degrees Celsius. Although its interior is warm, the heat does not radiate from its interior. It seems to be trapped by the atmospheric layers.

There are seventeen satellites of Uranus that have been discovered so far. It also has a ring system consisting of nine well-defined rings, plus a faint ring and a fuzzy one. Most of Uranus's rings are narrow, unlike those of Saturn, which are much broader. Like those of Saturn, the Uranian rings are composed mainly of rocky material ranging in size from dust particles to big boulders.

Neptune: Neptune is named after the god of the sea—identified with the Greek god Poseidon. Neptune is a gaseous planet with a diameter of about 50,000 kilometers (31,000 miles). Its average distance from the sun is about 4.5 billion kilometers (2.8 billion miles). It orbits around the sun in about 165 years and rotates about its own axis in about sixteen hours.

The atmosphere of Neptune is similar to its nearest twin, Uranus. It is about 80 percent hydrogen, 15 percent helium, and 3 percent methane, plus traces of other gases. The planet has a rocky core surrounded by a mantle of ice, followed by liquid molecular hydrogen. Its average surface temperature is -216 degrees Celsius. Its surface gravity is about the same as that of Earth. The presence of methane and icy particles in the atmosphere gives Neptune its blue color. Unlike Uranus, Neptune radiates out its internal heat. Its atmosphere is turbulent and dynamic.

Pluto: Pluto is named after the god of the underworld—also called Hades by the Greeks. As mentioned earlier, Pluto is currently classified as a dwarf planet.[2] Its diameter is about 2,300 kilometers (1,429 miles). Unlike its big gaseous neighbors, Pluto is tiny and rocky in composition.

Pluto has a long, extremely elliptical orbit around the sun that takes about 249 years to complete. It makes a complete rotation around its axis in about 6.4 days.

Pluto's interior is mostly made up of rock. Its surface is covered with methane ice, nitrogen ice, and carbon monoxide ice. Because of these ices, Pluto is highly reflective. The average surface temperature is -230 degrees Celsius. Its atmosphere is mostly made up of nitrogen gas with traces of carbon monoxide, methane, and other gases. It has a thin atmosphere with a pressure of less than 1/100,000 of Earth's atmospheric pressure at sea level.

Pluto has a large moon Charon, named after the mythical boatman who ferried souls of the dead across the river Styx to Hades. Charon has a diameter of about 1,200 kilometers (746 miles), which is nearly half the size of Pluto. Not very much is known about the interior composition of Charon. Its surface is covered with water ice. There is no evidence of any atmosphere surrounding Charon.

Asteroid: An asteroid is a rocky body with a size ranging from one kilometer (0.62 mile) to one thousand kilometers (620 miles) in diameter that orbits the sun. More than eight thousand asteroids have been catalogued in our solar system, and most of them lie in the asteroid belt between Mars and Jupiter.

Spectroscopic analysis of light reflected from asteroids shows that the composition of asteroids varies, depending on their distance from the sun. The inner orbit asteroids are made of mostly silicate rocks, while the outer ones contain water ice and carbon monoxide ice mixed with the rock. Asteroids are not known to have any atmosphere.

Comets: Comets are small icy bodies that orbit the sun in long eccentric orbits. Most of them are a few kilometers in diameter, but their tails can extend to millions of kilometers.

A comet consists of three parts: the nucleus, the coma, and the tail. The nucleus is a solid mass of ice that also contains dust or bits of rock ranging in size from minute particles to boulders a few meters across. Most comets have nuclei ranging from one to ten kilometers but some can reach diameters of over one hundred kilometers.

As a comet in its orbit periodically moves closer to the sun, the heat causes the nucleus ice to evaporate and turn into gas. This gas and dust surrounding the nucleus is called coma. The coma can reach more than a million kilometers across.

The tail of a comet consists of light dust grains that have been pushed out of the coma, away from the sun. The dust trail can reach a length of ten million kilometers or more. As the dust particles reflect sunlight, the tail appears to be white or yellowish.

The comets also have a gas or ion tail that consists of ions (atoms stripped of their electrons) trailing away from the sun by the action of the solar wind. These ion tails are bluish in color and can extend over one hundred million kilometers in length.

Meteors: In addition to the planets, moons, asteroids, and comets, there are small objects orbiting the sun. These are called meteors, meteorites, or meteoroids.

Meteors are commonly known as shooting stars. A meteor is a small piece of rock traveling through space at high speeds. If it enters Earth's atmosphere, it burns up and appears as a streak of light.

A *meteorite* is a large-size meteor that does not completely burn up in Earth's atmosphere and the unburned part hits the ground. The term *meteoroid* is a collective name for meteoritic bodies.

Dark Matter and Energy

The contents of the universe are matter and energy that exist in certain proportions at this time of the universe's evolutionary history. Matter is represented by over one hundred billion galaxies, intergalactic gas and dust, and a cosmic sea of particles that crisscross the universe. The universe is likewise bathed with radiation and energy fields. It is estimated that 4 percent of the total content of the universe is *ordinary matter*, 22 percent is *dark matter*, and 74 percent is *dark energy*.

The so-called "dark matter" cannot be seen like ordinary matter, but its presence is deduced from the rotation of galaxies and galaxy clusters. The missing mass of the dark matter, although undetected and unobserved, must exist to account for the gravitational field necessary to provide the dynamics of motion and organization of galaxies.

Dark energy is even more mysterious than dark matter. It is implicated in the expansion of the universe by playing the role of a mysterious antigravity substance. It is thought to be a vestige of a fierce force that drove the rapid expansion of the early universe (inflation). Although we do not know much about the nature of dark energy, it may hold the key to our understanding of the early universe and the nature of space itself.

Fate of the Universe

The word *universe* conjures up the image of an infinitely vast space that surrounds us and surrounds all the things that are contained in it, such as planets, stars, galaxies, and so forth. A precise definition of the term is difficult because we do not know the full extent of it. At best, we can call it the space-time continuum containing all the matter and energy that exists.

Not only is the true extent of our universe not known, there is a disagreement over whether the universe is finite or infinite in volume. However, this does not deter scientists from estimating the size of the visible or observable universe—the part of the universe that we can observe with our scientific instruments.

From the redshift data of the expanding universe (Hubble's Law) and the cosmic microwave background radiation (CMBR) measurements by NASA's Wilkinson Microwave Anisotropy Probe (WMAP), the age of the universe has been estimated to be 13.7 billion years with an uncertainty of 200 million years. That means that the region of space from which light (electromagnetic radiation) had time to reach us was about 13.7 billion light years away from us when the light started from it.[3] The present distance to the edge of the visible universe is much larger because the universe has been expanding since then. By taking this expansion into account, the present distance is estimated to be about 78 billion light years.

Although Earth is not the center of the universe and neither is any other point, planet, or star, we can define our visible universe as an imaginary sphere of about 78 billion light years radius, with ourselves at the center. The diameter or width of the universe is then at least 156 billion light years, which equals about 1.5×10^{24} kilometers or 0.9 trillion trillion miles.

Although we have been able to calculate the diameter of our visible universe by assuming it to be spherical, the shape of the universe is an open question in cosmology.

Is the universe closed, flat, or open? Will it continue to expand forever, or will it eventually stop expanding? If it stops expanding, will that be its final fate or will it collapse back on itself? Cosmology does not have a definitive answer with regard to these questions. However, a cosmological density parameter, called *omega (Ω)*, has been identified that will have a bearing on these issues.

Density parameter, Ω: The cosmological density parameter, Ω, is defined as the ratio of average density of the universe to the *critical density*. At the critical density, when $\Omega = 1$, the universe expands, but it slows down to a stop at an infinite time. In other words, the velocity of expansion approaches zero when the age of the universe is infinite.

If Ω exceeds one, it is a *closed universe* in which gravity dominates the force of expansion and eventually will cause the universe to collapse. The presence of excess matter/energy will curve the space-time continuum, as predicted by Einstein's general theory of relativity.

A universe with Ω less than one is an *open universe* that will expand forever. Because of its less matter (less gravity), it will have a negative curvature, like a saddle.

Between a curved universe and a saddle-shaped universe is the case of a *flat universe*. Such a universe has no curvature, positive or negative. It has a critical density ($\Omega = 1$) and, therefore, it will keep on expanding until it slows down to a stop at infinite time. Thus, the density parameter, Ω, is not only related to the expansion dynamics of the universe but also its shape. It is an important parameter that governs the evolution of our universe as well as its fate.

The critical density of the universe was introduced as a parameter in the general equations developed by Alexander Friedman in the early 1920s to model the expansion dynamics of the universe in accordance with the general relativity theory. A simplified relationship[4] for the determination of critical density can be derived by equating kinetic energy of the receding galaxies using Hubble's law of expansion and the galaxies' potential energy due to gravity.

Critical density involves Hubble's constant and the gravitational constant. A critical density of approximately 10^{-26} kg/m^3 has been calculated based on the best value of Hubble's constant available at this time. It should be noted that this is a very low density indeed, being on the order of six hydrogen atoms per cubic meter of space.

The calculation of actual average density of the universe is marred with too much estimation. Astronomers have added up the masses of all the known stars and galaxies and divided the total mass by the estimated volume of the visible universe to get the average density. It comes out to be less than the critical density.

Best estimate of Ω due to all atoms in the universe is 0.04 and about 0.22 due to dark matter (invisible or undetected). The latter is hypothesized to account for rotation and stable organization of galaxies and galaxy clusters. There must be matter, although undetected, to keep these galaxies rotating in place instead of flying off. Some possible forms of dark matter include interstellar planets, brown dwarfs, undetected neutron stars, black holes, massive neutrinos, and exotic particles called *WIMPS* (weakly interacting massive particles), which have been theoretically predicted but yet not observed.

From the above discussion, it appears that both the visible and dark matter adds up to only about 26 percent of the critical density. The other 74 percent of the critical density is thought to be contributed by dark energy, thus making our universe attain a density equal to the critical density ($\Omega = 1$).

Some theoretical evidence for dark energy comes from the observed recession speed of supernovae, which has been explained by the difference between the gravitational pull of ordinary matter and the gravitational push of dark energy as predicted by Einstein's cosmological constant. If this hypothesis is correct, the missing 74 percent of the mass/energy in the universe is supplied by this dark energy presumably spread throughout space.

The issue of whether the universe has a matter/energy density equal to the critical density is not settled definitively. However, the most refined data, theoretical as well as observational, favor the hypothesis that our universe has 100 percent of the critical density. Consequently, it is thought to be a flat universe with no curvature of its space-time continuum except for some wrinkles caused locally by the massive stars, galaxy clusters, and black holes.

What is the ultimate fate of the universe? The answer depends largely on its matter/energy density or Ω. If Ω is greater than one, the universe will eventually be overtaken by gravity. It will stop and then collapse back on itself in what is called a big crunch. The big crunch is the reverse of the big bang. What happens after the big crunch? It could be nothing or a start of another big bang; who knows? One thing is certain in such an eventuality, though: entropy of the closed universe will increase, so the next big bang will start with greater entropy.

In the case of an open universe (the most probable scenario), with Ω equal to or less than one, it will continue to expand, and the galaxies will get farther and farther apart. The galaxies will eventually deplete their star-forming gas and dust, and the old stars will run out of their fuel, turning into white dwarfs, neutron stars, or black holes, depending on their masses. For example, our sun will reach the end of its nuclear fuel in about five billion years from now. It will first swell up into a red giant, swallowing up Earth and other nearby planets. It will eventually contract to become a white dwarf, with a diameter of a few thousand miles.

After about ten billion years or so, when most of the stars in the universe have burned out, the gravitationally bound matter will eventually end up in black holes. Galaxies will dissolve, sink toward the center, and coalesce into massive black holes.

After about 10^{22} years from now, all matter outside the black holes will come apart. Even the most stable particle, the proton, will disintegrate into electron, positron, and neutrino.

Black holes evaporate slowly by emitting particles and gamma rays. This process is called Hawking radiation. It could take 10^{90} years for all black holes to evaporate away and disappear completely.

> This is the way the world ends
> This is the way the world ends
> This is the way the world ends
> Not with a bang but a whimper.[5]

CHAPTER 8
BIOLOGIC EVOLUTION

Primordial Earth

Our universe came into existence about 13.7 billion years ago. From a tiny point of energy, it grew into a universe of almost infinite dimensions. Matter formed out of energy and condensed into billions of galaxies, each containing billions of stars. One of these galaxies, in which we live, formed about twelve billion years ago. It is a rotating spiral-shaped disk of stars, about 30,000 light years (1.8 thousand trillion miles) in diameter and containing about 300–400 billion stars. It is our *Milky Way*.

Our solar system is located within one of the spiral arms of the Milky Way, about two-thirds of the way between its center and the edge. It was formed about 4.6 billion years ago when a rotating nebula of gas and dust collapsed on its center due to intense gravity. The pressure and temperature increased to the point of nuclear fusion, thus creating a star, our sun. Then the surrounding regions of the nebular cloud gradually cooled down and the particles began to condense into aggregate bodies, or planets, through the action of rotation, gravity, and accretion. Analysis of rocks on Earth, the moon, and meteors indicate that the sun, planets, moons, asteroids, comets, and other space debris within the solar system were all formed at the same time—about 4.6 billion years ago.

Although it is difficult to accurately reconstruct the early history of Earth, the following description is based on extrapolation of the current geological data and the environmental conditions that existed in the solar nebula. When Earth was formed, it was a hot body with a temperature of about two thousand degrees Kelvin. The original dust and gas got vaporized. As Earth cooled, the water vapor condensed. The interior of Earth contained mostly iron and silicates, with small amounts of other elements, some of them radioactive. As time passed, the heat generated by the energy released from radioactive decay melted the iron that settled in the core. A rocky mantle and crust formed around the core.

During the first half-billion years of its existence, Earth experienced violent volcanoes, intense solar ultraviolet radiation, fierce thunder and lightning, and a heavy bombardment of meteorites and comets. Its surface got pitted and cracked as it gradually cooled, letting steam and gases escape. Oceans were formed when the steam from the cracks and the volcanic eruptions condensed and gathered in the natural basins created by depressions in the crust. Melting of ice in the comets that impacted Earth also made a contribution.

The escape of gases from the earth's interior, as well as gases released from comets and meteorites, formed Earth's early atmosphere. It has since been modified by biological events, volcanic eruptions, and the incoming debris from space.

Origin of Life

When Earth was formed, it was a lifeless planet like the others in the solar system. But it orbited the sun at the right distance to support life. Planets closer in would be too hot, and those farther out would be too cold. The elemental composition of Earth—its oceans of water, its atmosphere, and its distance from the sun—provided the right environment for the atoms to bond together in molecular chains that became the ingredients of what we call life.

Earth's early atmosphere contained water vapor, nitrogen, carbon dioxide, hydrogen, and carbon. Spontaneous reactions between these atoms and molecules added methane, ammonia, and water to the mix. Additional chemical reactions took place that gave rise to more complex molecules by the catalytic action of energy provided by frequent lightning and other sources. Among these molecules were amino acids[1] and other compounds essential to life.

The idea that the initial conditions on Earth created the organic molecules that are the building blocks of life has been tested in the laboratory. In 1953, Stanley Miller, of the University of Chicago, conducted experiments to simulate conditions that might have produced complex organic molecules on the early earth. He exposed a gaseous mixture of hydrogen, ammonia, and methane in a vessel containing water to electrical sparks for several days. At the end of the experiment, he found a reddish brown residue rich in amino acids.

Similar experiments have been performed with water vapor, carbon dioxide, and nitrogen—now believed to be the constituents of early Earth's atmosphere—with the same results. Scientists have been able to produce all twenty of the amino acids, the building blocks of proteins, as well as purine and pyrimidine bases, the building blocks of nucleic acids like RNA and DNA and a number of other molecules essential to life.

Although further steps are needed to create actual life, these experiments demonstrate that life could have been created on Earth from chemical reactions between atoms and molecules present in its early atmosphere and oceans.

It is recognized beyond doubt that all living organisms are made up of atoms—the same atoms that constitute matter or any material object. What makes something into living, as opposed to nonliving, is the particular organization and bonding of specific atoms. The processes that govern the living follow the same laws of physics and chemistry as those that govern the nonliving. The difference is only in the expression of the process, not in its very nature.

This brings us to the question of what constitutes life. Because living organisms exhibit very diverse properties, it is difficult to define life precisely. However, some simple definitions have been devised that characterize most organisms. One of them is based on seven "signs of life," namely, *reproduction, growth, nutrition, respiration, excretion, senses,* and *locomotion.* But there are exceptions to these criteria. For example, not all plants have the ability of locomotion, and it is questionable whether viruses actually respire.

Among the many theories about how organic molecules were stringed together to form the first living cells is the idea that the individual molecules (monomers) were absorbed and aligned in some solid medium such as clay to concentrate these compounds. Under the influence of heat and ultraviolet light, these compounds could have randomly combined to make larger molecules such as polypeptides (precursors of proteins) and polynucleotides (precursors of DNA).

The next step in the development of life involved the formation of coacervate droplets—primitive cell membranes—to create sheltered enclosures for macromolecules to interact and produce complex chemical reactions while selectively allowing certain molecules to pass through.

Through millions of years of complex reactions taking place in enclosures—the coacervates—a point was reached when self-replicating molecules such as the DNA and RNA were formed.

The appearance of self-replicating molecules, equipped with primitive cell membranes, provided a selective advantage to these molecules to make the best use of raw material and energy resources available. They multiplied in great numbers, producing generations of prebionts, containing genetic information of the parent molecules.

Evolution of prebionts resulted in single-celled organisms, called *prokaryotes.* These primitive living entities had the outer cell membrane but lacked the inner membrane to enclose the nucleus of genetic material. They were much like the present-day bacteria.

Prokaryotes: These are the first cells of life to develop on Earth, some three billion or more years ago. They were much like today's bacteria and blue green algae. According to the microfossil records, they were the dominant form of life through the Precambrian era of Earth's life, 3.5–0.8 billion years ago.

Prokaryotes were cells without nuclei. They lived in water or on marine sediments. Since there was little or no oxygen in Earth's atmosphere then, these organisms were anaerobic—they functioned without oxygen.

For the two billion years or so of prokaryotes' abundance, the evolutionary change was relatively slow until the blue green algae began to produce oxygen through photosynthesis. Since oxygen is a highly reactive gas and dangerous to anaerobic organisms, it created a stimulus for cellular evolution. Rising oxygen levels triggered the evolution of prokaryotic cells to *eukaryotic* cells.

Eukaryotes: Eukaryotic organisms first appeared around 1.5 billion years ago. Unlike their ancestors, prokaryotes, eukaryotes possessed inner membranes to enclose the genetic material. Outside the nucleus in the cytoplasm, they also contained sophisticated organelles such as mitochondria for aerobic respiration and chloroplasts to perform photosynthesis. Eventually, the eukaryotic cells developed into colonies and gave rise to multicellular plants and animals.

Genetic Code

Reproduction is the essential criterion of life, whether it is a single-cell bacteria or a multicellular organism. In the reproductive process, the organism passes on some of its characteristics to the offspring. This is accomplished by a genetic code contained in a molecule called the *DNA* (deoxyribonucleic acid).

The DNA molecule, discovered by James Watson and Francis Crick in 1953, is an immensely long, double-stranded molecule with a helical arrangement. Each strand consists of a backbone of alternating sugars, called deoxyribose, and phosphate subunits. The two helical strands are linked together by nitrogen containing subunits called nucleotide bases. It is an amazing fact that in the DNA of all organisms, there are only four different types of nucleotide bases. They are *adenine (A)*, *thymine (T)*, *cytosine (C)*, and *guanine (G)*. These nucleotides act as letters from which the genetic language is constructed.

The "words" of the genetic language consist of the letters (e.g., AAA, TCC, CGT, and so on). A complete sequence of these bases, arranged in triplets, is the genetic code. The code contains instructions for manufacturing a specific amino acid. Amino acids link together to form proteins, and proteins make up the bodies of organisms as well as other necessary compounds such as enzymes and hormones. Thus, the genetic code contains all the information necessary to reconstruct a new organism.

It should be noted that from the sequence of four nucleotide bases of DNA (A, T, C, and G), exactly sixty-four different three-letter code words are possible. This mathematical fact holds true universally for all living organisms.

The three-letter words, put together to form codes of information for making specific proteins, are transcribed into *RNA (ribonucleic acid)*. The RNA is a close relative of DNA and comes in three different kinds—messenger, ribosomal, and transfer.

The *messenger RNA (mRNA)* is formed as a working copy of the DNA code when protein production is intended. After transcription, mRNA leaves the nucleus via pores of the nuclear membrane and goes into the cytoplasm of the cells. There, it interacts with the *ribosomes,* composed of *ribosomal RNA (rRNA)* and proteins, and directs the formation of specific proteins from amino acids provided by the *transfer RNA (tRNA).* The rRNA translates the genetic code carried by the mRNA, ensuring that correct amino acids are bound together to form the desired protein.

DNA is a very lengthy molecule (about ten feet, if unraveled and stretched into a straight line), but it is coiled up many, many times into a tight package called the *chromosome.* A human cell contains forty-six chromosomes in its nucleus. Of these, twenty-three come from the sperm of the father and twenty-three from the egg of the mother. The chromosomal number in a nucleus is called a diploid number, and the half set contributed by each parent is called the haploid number.

The diploid number varies greatly between species and does not bear a simple relationship with the complexity of the organisms. For example, the diploid number equals 46 in humans, 48 in chimpanzees, 254 in the hermit crab, 500 in some varieties of fern, and only 2 in the roundworm.

Ordinarily, chromosomes are tangled up in the cell and cannot be viewed individually. However, during the cell-division cycle, in the phase called mitosis, chromosomes copy themselves into two sets and can be seen easily. After the cell divides, the two new daughter cells contain identical chromosomes in their nuclei—the same as those of the parent cell. Thus, the same genetic information is passed on from a parent cell to the daughter cells after each division.

Time Scale of Life

Earth was lifeless when it was formed about 4.6 billion years ago. The earliest traces of microfossils date back to about 3.5 billion years. Because not all life forms were able to fossilize, it is thought that the earliest life on Earth predated the fossils by a considerable period.

The evolutionary history of life on Earth has been divided into eras, starting with the Precambrian. Each era is divided into periods, and the latest era, the Cenozoic, is further divided into epochs. These divisions follow the order and timing of some of the major events in the evolution of life on Earth. The dates for the eras and periods vary somewhat in the literature and are expected to change with new discoveries of fossils and other geological records. Here we will follow the dates from Philip Whitfield's book.[2]

Precambrian era, 4,600–570 million years ago (MYA): Life originated in the Precambrian era and consisted of simple cells without nuclei (prokaryotes). They included different forms of bacteria and simple plants resembling blue green algae. They lived in water and utilized the then prevalent carbon dioxide and sulfur in the atmosphere. The blue green algae gave off oxygen through the photosynthetic process, and the oxygen started building up slowly in the atmosphere.

Near the end of the Precambrian era, we begin to see fossils of clusters of cells with nuclei (eukaryotes) and multicellular organisms with sexual reproduction. There was no life on land.

Paleozoic era, 570–245 MYA: This era is divided into six periods. The Cambrian period (570–500 MYA) saw the flowering of animals (invertebrates) in the sea, such as trilobites, mollusks, and arthropods. Protozoa (single-celled eukaryotic animals such as amoebae, ciliates, and flagellates) dominated the water. Only algae represented the plant life. There was no life on land. The Ordovician period (500–440 MYA) saw abundance of marine animals such as trilobites and protozoa. The first fishes appeared. Among plants, bryophytes (e.g., mosses and liverworts) appeared on land. The Silurian period (440–410 MYA) saw fishes in abundance. Top predators were sea scorpions. Seedless vascular plants such as mosses and ferns were common. The Devonian period (410–365 MYA) saw amphibians come on land. Other land animals such as insects, crabs, and spiders also appeared. There was an abundance of seedless vascular plants on land and in swamps. The Carboniferous period (365–290 MYA) may be called the age of the amphibians. Insects evolved wings, and some amphibians evolved into reptiles. Seedless vascular forests dominated the land. The Permian period (290–245

MYA) saw amphibians decline, and reptiles began to dominate land animals. Conifers dominated plant life on land.

Mesozoic era, 245–65 MYA: This era is divided into three periods. The Triassic period (245–210 MYA) saw the first dinosaurs. Reptiles dominated the land; pterosaurs flew, and turtles and ichthyosaurs swam in the sea. The Jurassic period (210–140 MYA) saw the first mammals and birds. The first flowering plants (angiosperms) appeared. Gymnosperms (e.g., pines, spruces, cedars) dominated the land plants. Dinosaurs and insects dominated the land animals. The Cretaceous period (140–65 MYA) saw dinosaurs become extinct. The angiosperms spread and with them, the insects.

Cenozoic era, 65 MYA–present: This era in evolution is divided into two periods. The Tertiary period (65–2 MYA) saw dominance of land by mammals, flowers, and insects. Birds flourished. Bony fishes and zooplanktons dominated the seas. Around six million years ago, the evolutionary branch that led to *Homo sapiens* separated from the branch that led to the chimpanzees. The Quaternary period (2 MYA–present) is the age of the mammals. Early ape-man and humans evolved, leading to the most intelligent creatures on Earth, the modern human beings.

Theory of Evolution

The term *evolution* refers to a process of change over time. In the context of biologic evolution on Earth, evolution means the process that gave rise to the change of one species to another. A more technical definition describes evolution as any change in the frequency of one allele within a gene pool.

The theory of biologic evolution was conceived, independently, by British naturalists Charles Darwin and Alfred Wallace in the mid-nineteenth century. Their works were presented at the same meeting of the Linnean Society in London in 1858.

A detailed theory of evolution and the scientific data on which it was based was published in 1859 by Darwin in his book, *On the Origin of Species by Means of Natural Selection*. This extensively researched work provided a strong foundation for the idea that species change over time, some for the better and others for the worse. Those species for which the change is for the better can thrive, and the ones for which the change is for the worse can go extinct. There are shades of gray between the two outcomes.

Details of the evolution theory can be found in numerous books and articles on the Internet. Information presented here is extracted from these sources (see Bibliography).

All species, extinct or extant, are the result of evolution that began over three billion years ago with the single-celled organisms, the prokaryotes in the sea. While evolution worked to change the species, the process of natural selection acted to eliminate those that could not cope with the environment.

Darwin's theory of natural selection received a great boost in the mid-twentieth century when the genetic basis of heredity was applied to an evolving population of organisms. The original ideas of heredity were provided by the plant-breeding experiments of Gregory Mendel (1822–84). With the subsequent discovery of DNA and especially of its double-helical structure by Watson and Crick in 1953, Mendel's theory of genetics became a well-established science as well as an important component of the evolutionary theory.

The modern theory of evolution is a synthesis of Darwin's theory of natural selection and Mendel's theory of population genetics. Evolution can now be explained through mutation, genetic drift (genetic isolation of two populations), gene flow (genetic mixing), and natural selection.

The centerpiece of the evolutionary theory is that all life on Earth originated from a single point. In other words, all species have descended from a common ancestor—the progenitor organism. This aspect of evolution is called the *theory of common descent*.

Common descent: Evidence in support of the common descent theory may be found in the many traits that living organisms share universally. For example, all living organisms use the same genetic materials, amino acids, proteins, and cell-division processes for their metabolism and reproduction. They all exhibit the same signs of life: reproduction, growth, nutrition, respiration, excretion, senses, and locomotion. The universality of these traits at the molecular and functional level strongly suggests their common descent.

Discovery of prehistoric fossils and the accurate determination of their age through radiometric dating have allowed scientists to reconstruct the appearance, profile, and habitat of organisms in the context of geological times. Evolutionary trees have been constructed through careful analysis of fossils dating from the Precambrian era to the present Cenozoic era. These graphic representations of evolutionary data point to our common ancestry.

An important part of the fossil evidence is the continuity demonstrated by transitional fossils between two different lines of species such as the discovery of the *archaeopteryx* fossil. In some ways, this creature was like a bird with feathers and wings. In others, it was like a reptile with the body skeleton of a small dinosaur. It had a long, bony tail, claws at the front edges of the wings, and teeth in the jaws.

Further transition of reptiles into birds is seen in fossils of the Spanish and Chinese birds that showed shorter bony tails and smaller hand claws, as if the reptiles at this stage of evolution were becoming more and more birdlike.

Currently, there is an enormous amount of paleontological data available in support of the evolution theory. The data continues to grow with new discoveries of fossils that are filling the gaps in the evolutionary puzzle. It is amazing to see paleontologists make predictions for the missing links in the evolutionary tree and then find the fossils to fulfill the predictions. This is a familiar story in all fields of science. For example, many of the fundamental particles of matter were predicted theoretically before they were discovered experimentally.

Molecular evidence: One of the most definitive confirmations of evolution comes from molecular biology. It has been demonstrated that the degree of difference between DNA and RNA in different species is closely related to the time of the species splitting apart. For example, these molecules in humans are more identical to those in chimpanzees than in other vertebrates. Even among close relatives, human genes are more than 99 percent identical to those of the chimpanzees, slightly less for the gorillas and much less (about 80 percent) for the baboons. The evolutionary tree constructed on the basis of fossils and other evidence is, in fact, consistent with the predicted degree of molecular difference between the species.

Mutations: Genetic mutations are a fact of life. They are caused by changes in the genes that are part of the DNA. There are many known causes of genetic mutations: chemicals, radiation, heat, and other environmental conditions that can affect the genetic content of cells. Mutations can also be caused spontaneously by a so-called error in the duplication of genetic material during cell division. Although such a copying error is a rare phenomenon, it does occur as a matter of statistical probability. The frequency of error may be negligibly small, but it is not zero.

A genetic mutation by itself may or may not have a phenotypic effect. However, considering genetic recombination in the reproductive process, mutations can create new traits. Depending on whether the new traits provide an advantage or a disadvantage for the organism to cope with the environment, a gene mutation can become an agent of evolution. Although Darwin theorized evolution without the benefit of hereditary science, genetics has subsequently become the backbone of modern evolutionary theory.

Natural selection: Natural selection is a key evolutionary process whereby a genetic variant of a species establishes itself as a new species. The process is often called "the survival of the fittest." Being fit here means not only being adoptive to the environment but also having higher levels of reproductive success. These two traits, namely, adaptability and reproductive success, go hand in hand because variants that can better adapt to the environment are also fitter and produce more offspring.

Natural selection is not a deliberate act by nature to cause a demise of one species and install another in its place. Variation is a random event, but selection is governed by conditions in the environment. Success or failure of the variant cannot be predicted, and, therefore, evolution cannot be predicted.

The most obvious examples of natural selection are the instances in which organisms have developed characteristics that enhance their survival in a given environment. This is commonly observed among all forms of life—animals, plants, bacteria, and even viruses. For example, polar bears have evolved a white coat to camouflage against the snowy background so that their prey, the seals, cannot easily detect them. Trees have grown to be extra tall in thick forests to compete for sunlight. With time, insects become resistant to sprays, as do bacteria to antibiotics and viruses to antiviral drugs. In these instances, the organisms are not actively acquiring characteristics to enhance their survival. Rather, random mutations give rise to variants, and natural selection favors generations that have better chances of survival.

It should be realized that the time scale of natural selection in which a new species gets established is very long indeed. That is why natural selection cannot be observed at work in most cases. The story of biologic evolution and natural selection spans over 3.5 billion years. The process of evolution is stretched out so much in time that an observer at any given time would think of it as static—as if all the species of the world have been here forever. But the fossils tell a different story. It is like a long drawn out drama in which individual acts cannot be easily separated. In the words of Philip Whitfield: "Evolution resembles a great epic with a cast of millions."[3]

Scale of evolution: Evolution can take place to create a change below the level of species or above the level of species. The term *microevolution* is used to refer to an evolutionary change within a species population affecting the form or phenotype of organisms that make up that species. *Macroevolution*, on the other hand, refers to an evolutionary change that occurs at or above the level of species, resulting in the creation of a brand new species. There is no difference in the basic processes except that in the case of microevolution, the genes recombine and reshuffle within the species population; in macroevolution, the genes diverge into new species. Natural selection operates in both microevolution and macroevolution.

There has been some controversy with regard to theories that concern microevolution as a precursor to macroevolution; however, there is no denying the fact that macroevolution can happen by itself as a big genetic change to alter the species, provided it is favored to survive or thrive by natural selection. Such an occurrence may be rare, but evolution has a lot of time and opportunities on its hands. You may not be able to cause a macroevolution in the laboratory, but the earth's natural laboratory does not have limitations of time and resources to implement the laws of evolution.

The reason that microevolution is easier to accept than macroevolution is that the former can be demonstrated in the laboratory while the latter has to be inferred from fossil records and other indirect evidence, such as the traits of extant organisms and DNA comparisons.

Speciation: Macroevolution results in a new species. What is a species? It may be defined as a population of organisms in which individuals can breed with one another to produce fertile offspring but cannot do so with individuals of other species even when they have a chance.

Geographical isolation, or allopatry, can play a role in the formation of two or more species by creating distinctive gene pools through genetic drift or natural selection. The question of true species or subspecies is resolved if they come to occupy the same territory again, or sympatry. If the two subspecies reunite and fail to interbreed, that means speciation has occurred and they have become different species. On the other hand, if they can successfully interbreed, speciation has not occurred and genetic mixing would tend to homogenize the population.

Pace of evolution: The classical theory of evolution maintains that all evolutionary changes occur very gradually, in which microevolutionary changes blend into macroevolutionary changes. This theory has been challenged by some on the basis of fossil records (e.g., the fossils of freshwater snails and bivalves found in the Lake Turkana basin in East Africa) that indicate that the pace of evolution showed *punctuated equilibrium.*

The theory of punctuated equilibrium suggests that there are long periods of evolutionary stasis during which evolutionary changes take place, but interspersed therein are "punctuation events," characterized by dramatic evolutionary changes. The punctuated equilibrium theory is still not well established and remains to be further tested.

Human Evolution

According to the theory of evolution, all life, including humans, originated from the same progenitor organism—a single-celled prokaryote. The evolutionary tree is thought to have grown from a single point of origin and developed many limbs, branches, and leaves over a span of 3.5 billion years. It so happens that at the very top of this tree of life sits the most intelligent species Earth has ever known—the modern human being.

The understanding of human origins is derived largely from the analysis of fossil records, geological history, largescale migration of species, genetics, and the Darwinian theory of natural selection. Based on this carefully collected and analyzed data, the scientists have classified human beings, the *Homo sapiens*, as follows, per the Linnaean taxonomy classification system, originally constructed by Carolus Linnaeus (1707–1778):

- Domain: Eukaryota

- Kingdom: Animalia

- Phylum: Chordata

- Subphylum: Vertebrata

- Class: Mammalia

- Order: Primates

- Family: Homonidae

- Genus: *Homo*

- Species: *sapiens*

Although human evolution goes back to the very origin of life on Earth, we describe here man's evolutionary descent from the *primate* order.

Primates are thought to have evolved about seventy million years ago from shrew-like animals. The primate order split into three main lines about fifty million years ago: the prosimians (e.g., tarsiers and lemurs), New World monkeys, and Old World monkeys. A group called *hominoid* (e.g., apes and humans) then split off from the Old World monkey line. Further splits in the hominoid line took place between fifteen and five million years ago, separating into two families:

apes (e.g., orangutans, gibbons, gorillas, chimpanzees) and hominids (two-legged primates including humans and their immediate ancestors).

The earliest members of known hominids have been grouped into a genus called *Australopithecus*, now extinct, of apelike man who lived in Africa at least four million years ago. They could walk erectly and make tools. Their apelike cranium had an average capacity of 450 cubic centimeters (or cc), compared to modern man's average cranial capacity of about 1,350 cc.

The species of *Australopithecus*, *A. afarensis* and *africanus*, are considered ancestral in the lineage leading to the human genus *Homo*. The most famous members of *A. afarensis* are Lucy and Selam, whose skeletons were discovered in Ethiopia in 1974 and 2000, respectively.

The first-known member of the genus *Homo*, who lived in East Africa at least two million years ago, was *Homo habilis*. Evolution of *H. habilis* led to *Homo erectus* approximately 1.6 million years ago. The *Homo erectus's* bones and stone tools have been found in Asia, Europe, and Africa.

Homo erectus is thought to have evolved into *Homo sapiens* between 500,000 and 250,000 years ago. A subspecies of man, *Homo sapiens neanderthalesis*, evolved approximately two hundred thousand years ago from *H. erectus* or *H. sapiens*. This species had an average brain capacity of 1,600 cc, which is slightly larger than modern man.

Neanderthals lived mostly in Europe, although some lived in parts of Asia. They were displaced by *Homo sapiens*, who swept up from Africa about forty thousand years ago. There was an overlap of the two subspecies in Europe for about five thousand years. The Neanderthal man disappeared around thirty-two thousand years ago.

Transitional forms between *H. erectus* and *sapiens* are called archaic *H. sapiens*. The Neanderthal man could fall into this category, although some scientists consider it a separate species. Fossil records show that the transition to *H. sapiens* was complete by about 150,000 years ago in Asia and Africa and about 28,000 years ago in Europe. Modern humans are now the only surviving hominid species.

Anthropic Principle

The scientific theories of cosmic evolution and biologic evolution have been criticized because of their secularism—they do not presume any supernatural power that might have been responsible for the creation and evolution of the universe and life on Earth. As a matter of fact, science does not concern itself with any supernatural power just because it is outside its realm. Science, by definition, must follow the scientific method to find a natural truth. If anything is outside the preview of scientific methodology, it simply cannot be investigated by science. It is left to disciplines such as religion, spirituality, and philosophy to satisfy man's desire to understand what science cannot explain.

The anthropic principle and the theory of intelligent design or creationism are both outside the realm of science, although attempts have been made by some theologians, philosophers, and even scientists to provide a scientific basis for them. However, putting aside the question of a creator or God (which we will discuss in the next part of the book), these so-called theories do not meet the test of science. At best, they can be called pseudoscience. By saying this, I am not refuting the existence of a creator, an intelligent designer, or a God. I am just applying the scientific definition to these theories, and they evidently do not meet the criteria.

The anthropic principle simply states that the universe must be consistent with the origin and sustenance of carbon-based human life on Earth. It bases this assertion on the observation that all natural laws, physical constants, and the structure of the universe seem to be fine-tuned to the creation of intelligent life. Even an infinitesimal amount of deviation in any one of the critical parameters could have thwarted the very existence of the universe, let alone result in the emergence of life in it. For example, Stephen Hawking observed: "If the rate of expansion one second after the Big Bang had been smaller by even one part in a hundred thousand million million, the universe would have recollapsed before it ever reached its present size."[4] Paul Davies, another famous cosmologist, expresses similar sentiment: "The explosive vigor of the universe is thus matched with almost unbelievable accuracy to its gravitating power. The big bang was not evidently any old bang but an explosion of exquisitely arranged magnitude."[5] However, these statements should not be construed as being supportive of the anthropic principle. Scientists often marvel at the complexity, precision, and beauty of the universe and its natural laws. It has nothing to do with the existence or nonexistence of some supernatural power that might be responsible for the creation of the universe and its laws.

Brandon Carter first proposed the term anthropic principle in 1973 during a symposium celebrating Copernicus's five-hundredth birthday. In his presentation entitled "Large Number Coincidences and Anthropic Principle in Cosmology," he commented: "Although our situation is not necessarily central, it is inevitably privileged to some extent."[6]

In 1986, Barrow and Tipler published *The Anthropic Cosmological Principle*. In this book, they made a strong case for the anthropic principle by pointing out numerous instances in which it seems the universe had been perfectly set up for our existence. In other words, the universe has been designed deliberately with unbelievable precision to allow life to develop in it at some stage in its history. [7]

The reader is advised to read the above referenced books to judge the merit of the argument, but in my opinion, the whole concept is nonscientific and, honestly, a bit weird. Proponents of the anthropic principle in fact insinuate that the laws of physics were rigged in favor of the universe we have today, but no proof with any degree of scientific rigor has been put forth. At this point, it may be called a speculative philosophy at best.

Intelligent Design

The concept of intelligent design (ID) has the underpinning of the anthropic principle. It asserts that life on Earth, in all its details, was deliberately designed by one or more intelligent agents. It argues that the standard theory of evolution cannot explain the origin, diversity, and complexity of life. The fact that the universe is so fine-tuned for life means it must have been created by design.

Although ideas of ID have been around ever since Darwin proposed his theory of evolution, the concept of "intelligent design" has received considerable publicity since the publication of the book *Darwin on Trial* by Philip Johnson.[8] It has turned into an organized political campaign, very similar to its sister theory of creationism. In fact, ID and creationism are one and the same movement, designed primarily to discredit the scientific theory of evolution. It is unfortunate that this is being done through politics rather than scientific inquiry.

There is no scarcity of literature on ID and creationism. Numerous books have been written on the subject and the Internet can provide you with hundreds of matching sites if you care to Google anything even remotely related to the topics. My recommendation to the reader is: keep your head above the fog of irrational beliefs and unsubstantiated claims.

The key tenet of the ID movement is to impress on the public that scientists should not exclude supernatural explanations when naturalistic explanations fail to explain certain phenomena. The only problem with that is that the supernatural explanations cannot be subjected to an unbiased scrutiny. You either follow the scientific method, or it is no longer a science. There is simply no way to merge a supernatural view with the scientific view. So, instead of debating the issue at this point, I summarize below the salient features of ID, as a matter of general information:

1. Standard evolutionary models cannot account for the emergence of some very complex biological systems. These systems are "irreducibly complex," meaning that any attempt to dissect them analytically will cause them to cease functioning. The term *irreducible complexity* was coined by Michael Behe in his book, *Darwin's Black Box: The Biochemical Challenge to Evolution.*[9]

2. Living things are unique because they embody a large amount of "complex specified information"(CSI) that has a low probability of occurring by chance (a probability of 1 in 10^{150}). The argument of CSI was developed by William Dembski.[10]

3. A life-supporting universe is so fine-tuned that its random creation is exceedingly improbable. Thus, it must be a product of deliberate design.

4. Exclusion of supernatural possibilities by science amounts to an ideological bias that hinders the search for truth.

The anthropic principle and the concepts of intelligent design or creationism are elaborate attempts to inject supernaturalism into science or vice versa. Science has experienced these challenges before, and surely it is not going to buckle under any such pressures in this day and age. Fortunately, defiant scientists are no longer expected to face inquisition as in the past!

However, as scientists, we should not abstain from what comes naturally to man—a desire to understand who we are, where we come from, and where we are headed. In spite of our scientific theories of the universe and the life within it, we still do not know all the mysteries of nature. Our mind is limited in comprehending everything around us. We are constantly reminded of the possibility of a supreme being, a creator, who could have staged the whole thing. However, if we start mixing science with religion, we will contaminate both. It is better to keep them apart and not allow one to undermine the other. It is only through knowledge that we can reach a real nirvana—a state of perfect blessedness that comes from understanding the truth.

Accordingly, I recommend to the reader not to adulterate the scientific theories of cosmic and biologic evolution with the pseudoscientific notions of anthropic principle, intelligent design, or creationism. Take religion for what it is—a spiritual need and maybe an incomprehensible transcendental truth—but it is no science. We will try to address these issues and bridge the conflicts between the scientific and religious views after we have discussed religious thought in the next part of the book. In the meantime, let us stick with the scientific view and restate the basic principle of biologic evolution in the words of Philip Whitfield:

> The favouring of changes that are helpful in life's actual contexts seems to suggest that some guiding force is pushing evolution in a preordained direction. This is a recurring and seductive misconception. There is no external intelligence in life that guides evolutionary change. Evolution is not moving toward a predetermined goal nor is natural selection equipped with a compass. Since evolutionary change is born of random variation and local contingency, the only direction it can take is decided by those randomly generated changes which happen to be successful in specific circumstances.[11]

PART 3

FAITH

CHAPTER 9
RELIGIOUS PHENOMENON

I do not remember my first introduction to God as a dramatic event. In fact, it was not even an event or a special moment of discovery. I came to know God rather osmotically, just as I had assimilated other celestial ideas from my parents, family members, and village elders. My sporadic curiosity about God involved nothing more than asking few questions like how God made babies in their mothers' tummies, why the sun always rose in the east and set in the west, who caused thunder and lightning, and what constituted heaven, hell, angels, and ghosts. In fact, I accepted prevalent ideas in earnest, although mostly in awe, rather than conviction. About God and his manifestations, I never entertained any doubts, nor did I demand any proof. I took God and religion for granted—I was born with it, and it was a part of me.

Most believers also take their God and religion for granted. They acquire their faith from their parents or caretakers, and most continue to keep the same faith throughout their lives. As they grow older, however, their beliefs do get challenged on occasion by competing ideas, such as other religions, philosophies, and sciences. While most survive the challenge and continue to hold on to their childhood beliefs, some do succumb to the conflicts within or without. Religious conversions, although not common, do occur as a result of these conflicts.

Religious conversion is usually thought of as a change of the original faith into a different faith, such as a different sect of the same religion or a different religion altogether. Another type of conversion, although relatively less radical, is that of a person who makes a fresh start in the religion that he or she inherited. For example, a born-again Christian becomes more focused and dedicated to the faith than he or she had been in the past. Yet another kind of conversion is to abandon the concept of religion altogether, like becoming agnostic, atheist, or some other variety of nonbeliever. Human beings make these choices based on conditions and influences they are subjected to throughout their lives. Hopefully these decisions are based on intellectual reasoning and serious reflection rather than coercion or brainwashing by others. In any case, you are free to keep the religion of your parents or to follow a new path, barring any social, economic, or political pressures.

Psychology

The religious faithful usually abstain from subjecting their faith to scientific scrutiny. Most maintain that it is their intuition, feelings, or spirituality that connects them to God. Some call it simply a gift of God. They reject the thesis of those who characterize religious feelings and behavior as nothing more than a stretch of human imagination.

One reason that the religious faithful reject scientific analysis of their faith is that matters involving God and other related supernatural phenomena cannot be scientifically tested. They believe that their religious feelings emanate from their God-given spirit or soul, as distinguished from the material organs of the human body such as the brain.

Scientists do acknowledge the fact that proving or disproving of supernatural phenomena like spirit, soul, or God are beyond the scope of science. However, science can surely study the psychological, sociological, and anthropological aspects of the religious phenomenon. Such studies can gain insight into human emotions and thought processes underlying the religious beliefs. A few examples are listed below:

Ludwig Feuerbach: Ludwig Andreas Feuerbach (1804–1872) was a German philosopher who thought of religion as the outward projection of man's inward nature. He maintained that God corresponds in every aspect to human nature itself. Feuerbach did not believe in the reality of a God or gods. He believed that deities arose in the minds of human beings from their fears of the fearsome nature. Their worship and sacrifice were gestures to appease the fearsome. In essence, deities were nothing but the creation of the human mind.

Freudian psychology: Sigmund Freud (1856–1938) was quite blunt in his view of religion and considered it as another example of man's Oedipus complex. He traced the origin of religion to a primitive society scenario in which the father became jealous of his growing sons and drove them away at a certain age. The sons banded together, killed, and ate their father. But they also loved their father, and remembering the nurturing and protection he provided in their childhood created remorse. This feeling of remorse, combined with the guilt they felt about their sexual desire for their mother, created strong emotions of wrongdoing. They tried to neutralize these by substituting rites and moral codes, such as taboos against incest and endogamy. The guilt toward the slain father was compensated by making him into a god.

Freud not only compared religion to childhood neurosis, he thought that religion belonged to the infancy of the human race. In its transition from childhood to maturity, religion was needed to promote ethical values in order to build successful human societies. But now that humanity had come of age, it was no longer needed and should be let go.

Freud was quite antireligious in his outlook. He condemned religion because he thought it imposed restrictions on human choice and adaptation. He blamed religion for depressing the value of life and creating false beliefs and delusions. He considered religion as an impediment to man's mental and emotional development. For more of Freud's view of religion, read *Totem and Taboo* (1913) and *The Future of an Illusion* (1928), both by Sigmund Freud.

Jungian psychology: Carl Jung (1875–1961) added the concept of *collective unconscious* to the Freudian concepts of the conscious and unconscious. He suggested that religion is a collective response of human beings to their feelings of inadequacy and fears. Jung was more positive, however, toward religion's role in society. He considered it the expression of the instinctual forces of the psyche and the whole spiritual heritage of mankind's evolution. The collective unconscious motivates people to create myths, religious symbolism, and art, which are positive and healthy aspects of a human society. However, he did not rule out the pathological states that could give rise to collective neuroses.

Cognitive psychological approach: One of the important current hypotheses is that religious thought and behavior are not extraneous to the human brain. Pascal Boyer, an anthropologist and psychologist, takes a direct cognitive psychological approach in explaining the phenomenon of religion. He makes a strong case of the fact that religious thought and behavior are by-products of brain function.[1]

There is no doubt that the human mind is universally capable of generating religious feelings and supernatural imagination, but it is also incapable of detecting or measuring any supernatural phenomenon in the physical sense. That is why science is helpless in explaining the reality of supernatural phenomena that govern most religions. However, it can explain the physical processes involved in the acquisition of religious beliefs. For example, a cognitive psychological approach shows that religious concepts and beliefs are genuine and common to all human groups. They are a predictable by-product of ordinary cognitive function.

Sociology

Sociology is a branch of science that deals with human society, its social relations, organizations, beliefs, values, and so on. The sociology of religion pertains specifically to the study of religion in a social setting as well as its history and role in society.

The term *sociology* was coined by Auguste Comte (1798–1857), a French philosopher and social thinker. He hypothesized that religion is part of social evolution that human beings go through. The evolutionary process goes through several stages: believing in supernatural beings and forces, developing social laws and moral codes through the all-powerful divine authority, and modifying them through scientific exploration and debate to reach the practical social laws and governmental structures. Religion, in the evolutionary process, is the earliest primitive stage whereby people develop social codes of what is morally right and wrong in a civilized society.

Several sociological theories of religion have been developed over the last 150 years. We will briefly discuss only a few as examples.

Functionalism: The theory of functionalism was formulated by Emile Durkheim (1858–1917). The basic tenet of this theory is that religion is a social phenomenon and therefore should be explainable at the social level. Durkheim considered religion as glue that binds individuals to the society and as lubricant that facilitates smooth functioning of society by providing legitimacy, authority, social structure, and moral order.

The principles of Durkheim's functionalism theory are explained in his book, *The Elementary Forms of the Religious Life* (1912). By dividing the world into the sacred and the profane, religion is able to provide a moral code for the social functioning of a society. Rituals emphasize the importance of society to the individuals and reinforce society's moral authority. According to Durkheim, the idea of a God or gods is actually a secondary phenomenon. In other words, by worshiping a God or gods, human beings are in effect reaffirming their commitment to society.

The theory of functionalism has been elaborated by others to include human psychological needs, such as safeguards against fear of death and a feeling of control over their destiny.

Structuralism: The theory of structuralism was proposed by Claude Levi-Strauss (1908–), a French anthropologist. According to this theory, all social phenomena, including religion, originate from the innate structures of the human mind. Religious rituals, laws, and myths are revelations of the workings of the human mind.

The structuralists consider "facts" not as absolute facts but rather as interpretations by the preset structures of the human mind. The structuralism theory rejects empiricist leanings of functionalists for whom the scientifically verified facts are facts and established reality is reality.

Levi-Strauss maintained that there was no basic difference between the thinking of the so-called primitive human mind and the modern human mind. He insisted that the savage mind is equal to the civilized mind. He also believed that human beings did not have a privileged status in the universe. Like other species, they leave a few traces of their passage in time before becoming extinct.

Historical interpretation: Max Weber (1864–1920) introduced the concept of *historical interpretation* of religion. He identified three stages in the evolution of religion: (1) magic, (2) animism, and (3) monotheism.

The first stage in the evolution of religion involved magic. A magician was believed to be endowed with special powers to control natural phenomena such as rain, earthquakes, health, and fertility. The second stage produced animism—a belief that all life is produced and controlled by spiritual forces, separate from matter. Animistic beliefs gave rise to the ideas of gods and spirits. These concepts became ritualized by religious professionals: priests and prophets. The priests took over the job of rationalization, organization, and dissemination of religious ideas. The prophets fortified their ideas by claiming the authority of personal revelation.

The third stage, monotheism, involved the concept of a single God—a much more powerful deity, by virtue of the concentration of the powers of all gods and spirits into one entity. At the same time, the concept of the one supreme being eliminated the element of magic from religious beliefs. All supernatural phenomena now belonged to one God and not any god, spirit, ghost, or human claiming such powers. This elevated the rationalization of religion to a higher level.

Socioeconomic view: Karl Marx (1818–1883), a great social thinker of his time, had a dim view of religion. He considered religion to be a hindrance to reason. Because of that, he held the view that religion masked the truth and misguided its followers.

Marx's view of religion is best summed up by his famous "opiate" quotation: "Religion is the sigh of the oppressed creature, the heart of a heartless world, and the soul of soulless conditions. It is the opium of the people."[2]

Marx's major work concerned the social, economic, and political ills of the society in which he lived. His theories concerning these issues laid the foundation of political systems such as socialism and communism. These systems were, and they still are, in conflict with the economic system called capitalism. It appeared to Marx that religion pacified people with illusory rewards, thus preventing them from rising against capitalism. Instead of fighting inequality between the upper (capitalists) and the lower (laborers) classes, religion favored acceptance of the status quo between them. Instead of waging a struggle against capitalism, Marx believed religion offered solutions that were either irrelevant to the economic condition of the masses or concerned only with their utopian paradise. Marx's appraisal of religion was straightforward: it "eased pain even as it created fantasies" when dealing with the suffering of the poor at the hands of the rich and powerful.

Theory of formation: A general sociological theory of the formation of religions has been proposed by Rodney Stark and William Bainbridge in their book, *A Theory of Religion*.[3] According to this theory, most religions start out as cults or sects. A cult refers to a group that has developed beliefs and rituals so different from the prevailing norms of the society that it creates a high degree of tension. Many cults do not survive the societal tension or conflicts and eventually die out, but some manage to become more established and live with a certain degree of accommodation within the society.

Sects are formed the same way as cults, namely, by creating a novel theology, but sects differ in that they continue to remain within the basic edicts of the mainstream religion. A more established and autonomous group within a religion can reach the status of a denomination. Thus, a religion may have several denominations, and each denomination may have several sects.

A new religion comes into being due to a variety of reasons. Several sociological models have been proposed that apply in varying degrees to all religions. For example, a cult or a religion might have formed due to a period of great stress in the life of the founder. In other words, the founder suffered psychological problems and founded a new religion as a form of self-therapy. This is called the *psychopathological model*.

The *entrepreneurial model* offers another possibility of formation of a new religion in which the founder acts like an entrepreneur in developing a new religious product. The founder in this case makes use of ideas from the existing religions and improves on them to make them more appealing to people.

The *social model* involves a religious group developing an extreme level of affection and cohesion among its members. The social bonding is strengthened and carried to a level of what is called *social implosion*, leading to the formation of new theology. The new religion thus formed acts as a glue for the group and sustains its bonding.

The *normal revelation model* explains the formation of a new religion when the founder interprets natural phenomena as supernatural and ascribes them to a higher power or deity.

It should be noted that these sociological models are attempts to explain how groups and society have come up with new religions, cults, sects, and denominations. These theories are based on careful observation of cultures and societies, following scientific methodologies characteristic of social sciences. They do not incorporate in any way or form the supernatural elements of a religion, such as divine revelation or intervention, in the process. So, the reader is faced with this dichotomy at every step of the inquiry: a Godless scientific explanation versus unquestioning faith in the supernatural. There is absolutely no compromise between the two approaches to get at the truth of religion.

Anthropology

Anthropology is the study of man and his culture, social relationships, institutions, customs, and so on. The anthropology of religion specifically involves the study of religious institutions, beliefs, and practices across different cultures.

One area of interest in anthropology is the evolution of religion itself in relation to the biological and cultural evolution of humans. For example, some of the questions of interest are: Is there a correlation between religious evolution and cultural evolution? Can we call a religion "primitive" or "modern" depending on the state of social and cultural development? Are religious beliefs and practices reflections of political and economic forces (e.g., theories of Marx, Durkheim, and Weber)?

One problem with the anthropological study of religion is its stance on the truth of religious beliefs. Most anthropologists have difficulty in understanding them from the perspective of the believer. At best they tend to look at religious truths metaphorically rather than in the literal or scientific sense. Most, however, agree with Firth who said: "There is truth in every religion. But it is a human truth not a divine truth."[4]

As stated earlier, social studies of a religion do not address its central focus, namely, the belief in supernatural power(s). Investigation into the nature of divinity or similar phenomena is outside the realm of anthropology or any other discipline that is based on the scientific method of inquiry. However, anthropologists do regard religion as a "massive output of human enterprise"[5] and recognize its crucial role in society. In the words of William Howell: "Man's life is hard, very hard. And he knows it, poor soul;

That is the thing. He knows that he is forever confronted with the Four Horsemen—death, famine, disease, and the malice of other men."[6]

Theology

Religion has been studied in many ways. Some studies involve the scientific methods of psychology, sociology, and anthropology. Others involve philosophy in which religion is analyzed using logic, epistemology, metaphysics, ethics, morals, emotions, and so on. Philosophic studies, however, may or may not presume the truth of any religious belief.

Yet another way of researching religion is theology. It is the study of religion within the framework of its own principles, doctrines, and traditions. Theology assumes the truth of certain religious beliefs in order to promote rational understanding of religious structure, theories, doctrines, and concepts. The fact that theology has to rely on its presumption of certain religious dogma as given truths essentially distinguishes it from social sciences and philosophy, which try to minimize any preconceived notions in favor of or against the religion of interest.

Religions are mostly founded through religious movements that start out initially as collections of new ideas about what is holy or sacred. Some of these beliefs are later compiled into sacred writings called scriptures. As a religion spreads, further explanations and interpretations of its beliefs are added to the literature. At this stage, the organizational structure of the religion is rudimentary or nonexistent. It is only when and if the religion becomes significantly organized and established that its followers feel the need for its rational investigation and research.

A well-established religion gradually gives rise to theological exploration. Formal courses in theology are taught in various schools, universities, and seminaries. Currently, all major religions and denominations offer formalized religious teaching and research in an academic setting of some form or another.

It should be noted that theology is the study that relies on the theologian's own belief in a particular religion. Although it is not essential to be a believer in the religion you study, it is difficult, if not impossible, for a nonpractitioner to do rational investigation and testing of a religion. In other words, theological research is the province of those who are either followers, or at least sympathizers, of the religion. Because religion involves deep emotions and personal experiences rather than just intellectual thought, it is impossible to engage in theology with complete emotional detachment and professional neutrality. Thus, the first and foremost requirement for a student of theology is to be a firm believer. You have to believe in God before you can learn about God.

CHAPTER 10

ULTIMATE REALITY

The next two chapters are presented as a primer on world religions. The discussion focuses on the religious views of creation as well as the conceptual differences between major religions regarding who or what created the universe. This is in preparation for a dialogue between various faiths and science about the role of God as the creator of the universe in the later chapters.

The term "Ultimate Reality" has been used as a religious concept to signify a transcendent or immanent being (a personal God) or an eternal truth, principle, or process (an impersonal entity) that governs the universe. In all these descriptions, the Ultimate Reality is greater than the physical universe.

Although different religions conceptualize Ultimate Reality in different ways, it is the central belief or concern of all religions. In any religion, it seems the most important activity of human beings is to establish their relationship with what they consider as the Ultimate Reality.

The quest for God is in fact the quest for Ultimate Reality. Since the concept of such an ultimate truth or being is central to most religions, it is logical that my quest for God should lead me to consult major religions of the world for guidance, in addition to other sources such as science and philosophy.

There is no clear concept of a personal God in the Eastern religions such as Hinduism, Buddhism, and Taoism; rather the concept is that of the Ultimate Reality as an eternal truth or principle. In Western religions, namely, Judaism, Christianity, and Islam, the transcendent reality is God. The Ultimate Reality in their cases is believed to be the personal God who is immanent, omnipotent, and creator of the universe (and whatever is contained therein).

We start our quest for the Ultimate Reality by first learning how it is understood in the Eastern religions.

Hinduism

Hinduism is one of the oldest textually based religions of the world. It is not a single well-defined religion, but rather a conglomeration of many religious and philosophical thoughts that originated in India. Numerous cultures have contributed to its evolution. One of the earliest cultures that influenced Hinduism was the Indus Valley civilization that existed in northwest India some 4,500 years ago. This civilization flourished from about 2,500–1,500 BCE[1] and was centered on two cities, Mohenjo Daro and Harappa, located in present-day Pakistan. Archaeological excavations have revealed the existence of a mother goddess cult, which appears to be a forerunner of the Hindu goddess, Mahadevi. Rituals such as bathing to attain purification and the worship of animals date back to the same culture.

Around 1,500 BCE, there was a great migration of people who called themselves Aryans from the Caucasus region of central Asia to the Indus Valley of northwest India. They brought with them their culture and traditions. Through mixing with the existing culture of the Indus Valley, they gave rise to a new culture or religion called Hinduism.

Scriptures: As the Hindu culture in India further developed, religious ideas that had evolved through oral traditions gave way to the written tradition. It was between 1500–500 BCE that the religious beliefs of Hinduism were put into writing. The sacred texts of this period are known as the Vedas.

There are four Vedas (Rig-, Sama-, Yajur-, and Anthara-Veda) that constitute the oldest scriptures of Hinduism. They are mostly a collection of hymns, describing the works of various deities and praising their actions.

In addition to the above four Vedas, there are three other writings that were added later to the Vedic literature: the Brahmans, the Aranyakas, and the Upanishads. All these writings collectively are considered to be the most sacred scriptures of Hinduism. Two other important works of Hindu literature, called the Mahabharata and the Ramayana, were written between 300–100 BCE.

The Mahabharata is an epic poem containing about 200,000 lines. It is a war story interspersed with discussions of Hindu life and beliefs about nature, laws, politics, science, and creation. A major section of the Mahabharata is known as the Bhagavad Gita (the Song of the Lord). It is a dialogue between Arjuna, the warrior, and his charioteer, Krishna, who is the incarnation of the god Vishnu.

Ultimate Reality: According to the Upanishads, the Ultimate Reality is Brahman. Although Brahman is believed to be the Ultimate Reality, the nature of Brahman is not well understood. Some see Brahman as an impersonal absolute reality, the source of all things and all beings. In this concept everything is Brahman in the ultimate sense, but Brahman cannot be described in human terms.

There are others who view Brahman as a personal deity or a supreme God who is responsible for creating the universe and who appears to humans in various forms known as gods and goddesses. Therefore, you can communicate with Brahman through the worship of gods and goddesses because they are the manifestations of Brahman.

The chief Hindu gods are Brahma, Vishnu, and Shiva. Together, they are called the *Hindu Trinity* (not the same concept as the Christian Trinity). Brahma is the Lord of all creation. Vishnu is the preserver and is responsible for controlling human fate. Shiva is the destroyer and the Lord of time. This Trinity of gods represents the central belief of Hindus in the never-ending cycle of creation, preservation, and destruction.

Brahma is considered to be the most high and above human worship. He is depicted in paintings and carvings with four faces and four arms, often sitting on a swan or a lotus flower from which he was born.

Vishnu has appeared on Earth in a number of incarnations, one of them being born as Krishna, "the divine cowherd," who loved the beautiful mortal, Radha. The love and romance between a god and a mortal symbolizes the union of a human soul with divinity.

Another incarnation of Vishnu is Rama, the hero of one of the great Hindu epics, the Ramayana, which tells the story of Rama's wife, Sita, who was kidnapped by a demon, Ravana, and then rescued by Rama with the help of the monkey god, Hanuman. The triumph of god Rama over the demon Ravana is symbolic of the triumph of good over evil.

Shiva is responsible for destroying creation as well as recreating it. Shiva's wife, Kali, represents destruction of evil and ignorance. She is usually depicted as a fierce deity surrounded by decapitated heads and limbs. She also represents the benevolent character of Shiva and favors those who seek knowledge and help maintain the world order. Another form of Shiva's wife is Parvati, who represents kindness and gentleness. Shiva and Parvati had a son, the elephant-headed god Ganesh, the remover of obstacles. Ganesh is a very popular Hindu deity who is believed to be very fond of candy and is usually depicted with some candy in his hand.

Besides the gods mentioned above, there are many other gods in Hinduism. For example, gods from the Vedic period represent forces of nature and include Varuna, Indra, and Agni. Varuna is the oldest supreme god who maintains the cosmic order. Indra is the god of war, storms, and rain. Agni is the god of fire.

As already mentioned, the supreme god is Brahman, an impersonal force who created the universe. The gods are merely the manifestations of Brahman.

The Hindu concept of God embodies everything that the universe has to offer—creation, destruction, good, bad, violence, peace, beauty, ugliness, love, and hate. The source of all these phenomena—emotions and attributes—is one reality, the Ultimate Reality. One may call it God. The Hindu gods, Brahman, Indra, Brahma, Vishnu, Shiva, Krishna, and Rama, are all part of the Ultimate Reality. They are the embodiment of the entire universe, the animate and the inanimate. The Supreme God of Hinduism, therefore, whether Brahman or his incarnates, represents the oneness of all reality, which includes the nature of all things and beings—positive or negative. The God of the Western religions, on the other hand, represents only the good while attributing evil to another being called the Devil. Except for this conceptual separation of attributes, there is little difference between the Supreme God of Hinduism and the one God of the Western religions.

Cosmogony: The creation of the universe is closely linked to God and, therefore, is addressed by many religions. The Hindu scriptures describe the creation as follows:

> At first was neither Being or Non-being.
> There was not air nor yet sky beyond.
> What was its wrapping? Where? In whose protection?
> Was water there, unfathomable and deep?
> There was no death then, nor yet deathlessness;
> Of night or day there was not any sign.
> The One breathed without breath, by its own impulse.
> Other than that was nothing else at all.
> Darkness was there, all wrapped around by darkness,
> And all was water indiscriminate. Then
> That which was hidden by the void, that One, emerging,
> Stirring, through the power of ardor (tapas), came to be.[2]

Buddhism

Buddhism started in India in the fifth century BCE. Its founder was Siddhartha Gautama, who is known as the Buddha (the Enlightened One). He was born in what is now Nepal around 480 BCE and lived for eighty years.

Gautama renounced early on the life of luxury that his family had to offer and set out to find the meaning of life and the human condition. After a prolonged period of search and meditation, he achieved a state of enlightenment while seated under a tree, known as the bodhi tree, located near Gaya in northeast India. The enlightenment gave Buddha the knowledge and wisdom he had sought through years of contemplation, meditation, and inner struggle.

After achieving enlightenment, Buddha wandered about in northeast India with his band of disciple monks. His teachings recommended monastic life, which eventually gave rise to a stable monastic community called the *sangha*.

After about forty-five years of traveling around in India, teaching and living the life of a beggar, Buddha died at the age of 80. He comforted his followers with the following last words: "Do not cry. Have I not told you that it is the nature of all things, however dear they may be to us, that we must part with them and leave them?"[3]

Transcendent truth: Buddhism has no concept of a personal God. Rather, the concept is of Ultimate Reality as a process, an absolute truth underlying the state of being. This reality is transcendent and is devoid of all empirical determinations. It is not any kind of a primordial being such as Brahman, nor is it a personal God responsible for creating the universe.

Buddha was silent about the existence or nonexistence of God. In fact, he avoided the religious dogma and metaphysical speculation about a supernatural being. However, he did not reject the idea of gods. To him, gods represented beings who had attained their status by using the same virtues as any human could. In other words, gods may be thought of as human beings who had achieved enlightenment. In that sense, Buddha could also be called one of the gods. But he did not assign any supernatural powers to gods, including to himself. Accordingly, gods do not need to be worshipped nor should they be made the basis of a moral code. They do not have the power to make anybody happy or unhappy.

The concept of Ultimate Reality varies somewhat between the two major sects of Buddhism: the Theravada and the Mahayana. In the Theravada school, the Ultimate Reality is *nirvana* or *dharma*. Nirvana is the state of perfect blessedness in which all human desires, passions, hatreds, and delusions are extinguished. Dharma is the cosmic order or natural law including moral order, norms of behavior, and ethical rules that apply to all beings and things. Thus, the Theravada branch concentrates on the attainment of nirvana as the central goal of life.

In the Mahayana Buddhism, the concept of reality is incorporated in the three bodies, or trikaya, of Buddha: (1) the physical or earthly *body* of the founder with which the Buddha appears to humanity to provide guidance toward enlightenment; (2) the body of the *bodhisattva,* a semi-divine being who has achieved enlightenment but voluntarily puts off nirvana in order to help others to achieve enlightenment; and (3) the *dharma* body of the Buddha, or Dharmakaya, the highest and truest nature of the Buddha, identical with the Ultimate Reality.

The state of Dharmakaya or the Ultimate Reality, according to Mahayana Buddhism, is devoid of any permanent attributes and cannot be described by humans. It has been called by other names like "suchness" or "thatness." Because all things are empty of substance or enduring essence, this has led to the concept of *shunyata* (literally "emptiness" or "voidness") as the Ultimate Reality of the world. It is the void from which nothing escapes or is beyond it, not even nirvana. The word *Dharmakaya* identifies the dharma body of the Buddha with the Ultimate Reality. Thus, the goal of life according to the Mahayana school of Buddhism is to pursue enlightenment, but instead of achieving nirvana single-mindedly, it teaches involvement in the world to show love and compassion for others and to help them in the quest for nirvana.

Sikhism

The Sikh religion was founded by Guru Nanak in the AD 1400s in the Punjab province of India (presently divided between Pakistan and India). The religion grew out of the conflict between Hindus and Muslims of India and offered a conciliatory message by Guru Nanak: "There is neither Hindu nor Muslim. So whose path shall I follow? ... I shall follow the path of God."[4]

Sikhs believe in one God—the creator of the universe and whatever exists within it. The Supreme God is invisible, but his attributes are known through the wise and holy teachers called *gurus*. Starting with Guru Nanak, Sikhism has ten human gurus and a final guru that is the holy book, the Guru Granth Sahib. The message of the ten gurus and the scriptures is centered on God and the ways to attain his love through good works and service to humanity.

Taoism

Taoism was founded by a Chinese philosopher, poet, and sage, Lao-tzu (also spelled Lao Tse), around the sixth century BCE. Tao, which literally means "the way," emphasizes the life of moderation, avoiding extremes of any kind. It is believed that all living beings contain two complementary and opposite modalities of life: *yin* and *yang*. Yin is the dark, negative, and regressive nature, while yang signifies the light, positive, and progressive aspect of nature. It is the proportion and dynamic mixture of the two that determines the nature of living beings.

The Ultimate Reality in Taoism is the Tao—the cosmic laws that govern all phenomena in the universe. It is not a personal God or a being. It is the immutable and unchanging principle that underlies the creation of the universe and all beings. Human beings become moral when they identify themselves with this reality. This brings them into harmony with the universal laws.

The Tao represents the universal law as well as the Ultimate Reality. To be in tune with Tao is to be moral—in harmony with the universal law. Tao is therefore a prescription for moral life without invoking divinity.

Confucianism

Confucianism was founded by the Chinese philosopher and teacher, Confucius, around the sixth century BCE. Rather than a religion, Confucianism is an ethical system primarily concerned with moral conduct and harmonious relations with society and state. This development of morality and harmony occurs not only through physical actions but also through spirituality. Confucius believed that there was a higher power that guided humanity and he called it *heaven*. The heaven disapproved of chaos and approved of *harmony*. The balance between the complementary forces, the yin and the yang, was required to create this harmony.

The divine connection between heaven and humanity in Confucianism is not the same as that between a personal God and humanity. It is more like the concept of a higher power that had directed the universe and provided guidance for humanity. Confucianism is a code of moral conduct on earth, the principles of which are based on divine guidance. The emphasis here is on the search and implementation of moral code rather than the worship or appeasement of the divine power.

Shintoism

Shintoism is an ancient religion that originated in prehistoric Japan. It literally means "way of the gods." The Shinto beliefs are based on legends and mythology involving gods who are believed to control the natural elements such as thunder, rain, winds, mountains, streams, oceans, forests, and other objects and phenomena that influence human life on Earth. The supreme deity of Shintoism is Amaterase, the goddess of the sun. Until recent times, she was said to be a distant ancestor of the ruling imperial family of Japan.

The Shinto gods have great influence on their followers' lives. They are honored with their individual shrines and numerous rituals.

In addition to the gods, there are spirits, called kami, that are also honored to bring good luck, prosperity, and happiness. These spirits, if ignored, could cause terrible things to happen and therefore are approached with great respect and reverence.

Shintoism is practiced by most Japanese as a folk religion, often mixed with Buddhism, Christianity, and other religions. As Buddhism spread in Japan in the sixth century CE, it absorbed the existing beliefs instead of discarding them. For example, gods and spirits of Shintoism were combined with Buddhas and bodhisattvas. Currently most Japanese practice Shintoism mixed with varying degrees of Buddhism, Christianity, and other religions. Interreligious conflict is practically nonexistent.

The Shinto concept of Ultimate Reality may be inferred from its belief in spirits and divine guidance. Like the other Eastern religions, this belief is in the realization of a supreme power that controls the universe rather than a personal God who requires worship and prescribes reward and punishment for human behavior.

CHAPTER 11
GOD OF ABRAHAM

The three great monotheistic religions of the world, Judaism, Christianity, and Islam, owe their beginnings to a man, prophet, and patriarch named Abraham. Although numerous holy men and prophets preceded Abraham (starting with the first man Adam), it was Abraham to whom God made the promise:

> And I will make thee exceedingly fruitful, and I will make nations of thee, and kings shall come out of thee.
> And I will establish my covenant between me and thee and thy seed after thee in their generations for an everlasting covenant, to be a God unto thee, and thy seed after thee. (Gen. 17:6–7)

From the Bible we also learn that Abraham had two sons, Isaac and Ishmael. God promised Abraham that, in return for his obedience, he would bless his sons and make great nations out of them.

> And God said, Sarah thy wife shall bear thee a son indeed; and thou shalt call his name Isaac: and I will establish my covenant with him for an everlasting covenant, and with his seed after him.
> And as for Ishmael, I heard thee: Behold, I have blessed him, and will make him fruitful, and will multiply him exceedingly; twelve princes shall he beget, and I will make him a great nation. (Gen. 17:19–20)

From these stories as told in the Bible, God commanded Abraham to leave his home (Ur of the Chaldees—somewhere in present-day Iraq) and travel to a land called Israel. Generations followed Abraham, and his clan multiplied into twelve tribes. At some point in history, the twelve tribes of Israel were taken into slavery by the Egyptians and then liberated by Moses, who became the founder of the first monotheistic religion, Judaism. The word of God, the God of Abraham, was revealed to Moses on Mount Sinai (probably in the Sinai Peninsula in Egypt), some 3,500 years ago.

About 1,500 years later, a Jewish teacher named Jesus affirmed the God of Abraham as told in the Hebrew Bible (also known as the Old Testament). He founded the second monotheistic religion, Christianity, in which the authority of the Old Testament was accepted but superseded by God's message through Jesus, as recorded in the New Testament (the Christian Bible). Christians believe that Jesus Christ is the Son of God and that the Father and the Son are one and the same God—the God of Abraham. Jews, however, do not accept the divinity of Christ.

About 1,400 years ago (seventh century CE), the third monotheistic religion was founded by the prophet Muhammad, who received the word of God in Mecca (present-day Saudi Arabia). His message stressed the oneness of God (the God of Abraham). He affirmed God's revelations received by Moses and Jesus Christ, who were recognized as prophets of the one God.

Muhammad received the word of God through a series of revelations that are compiled into a book, the Qur'an. The Qur'an states very clearly and emphatically that God (called Allah in Arabic) is the one and the same God as the God of Abraham:

> You say: "We believe in Allah, and the revelation given to us, and to Abraham, Ishmael, Isaac, Jacob, and the Tribes, and that given to Moses and Jesus, and that given to (all) Prophets from their Lord: we make no difference between one and another of them: and we bow to Allah (in Islam)." (Qur'an 2:136)

It should be mentioned that while Jews, or Israelites, are considered to be the descendants of Abraham through Isaac, the Arabs trace their lineage from Abraham through Ishmael. If these claims and biblical stories are true, it seems God's covenant with Abraham was apparently fulfilled through the nations of Judaism, Christianity, and Islam. In spite of the theological differences between these religions, they surely have one thing in common: acknowledgement of the one God, the God of Abraham. Accordingly, the three religions, and others of the same creed, are often referred to as the Abrahamic faiths.

The question then arises: why did God allow more than one religion or nation to fulfill his covenant with Abraham? Only the Qur'an offers the explanation:

> To you We sent the Scriptures in Truth, confirming the scripture that came before it, and guarding it in safety; so judge between them by what Allah has revealed, and do not follow their vain desires, diverging from the Truth that has come to you. To each among you We have proscribed a Law and Open

Way. If Allah had so willed, he would have made you a single people, but (His plan is) to test you in what He has given you: so strive as in a race in all virtues. The goal of all is to Allah; it is He that will show you the truth of the matters in which you dispute. (Qur'an 5:48)

In more general terms, God said:

O mankind! We created you from a single (pair) of a male and a female, and made you into nations and tribes, that you may know each other (not that you may despise each other). Verily the most honored of you in the sight of Allah is (he who is) the most righteous of you. And Allah has full knowledge and is well acquainted (with all things). (Qur'an 49:13)

Judaism

Moses, the founder of Judaism, received God's commandments on Mount Sinai. The message was revealed to him through a series of conversations with God, who remained invisible but spoke to him out of a burning bush. The introduction went like this:

> Moreover he said, I am the God of thy father, the God of Abraham, the God of Isaac, and the God of Jacob. And Moses hid his face; for he was afraid to look upon God. (Exod. 3:6)
>
> When Moses asked God his name, he was thus told:
>
> And God said unto Moses, I AM THAT I AM: and he said, Thus shalt thou say unto the children of Israel, I AM hath sent me unto you.
>
> And God said moreover unto Moses, Thus shalt thou say unto the children of Israel, the LORD God of your fathers, the God of Abraham, the God of Isaac, and the God of Jacob hath sent me unto you: this is my name forever, and this is my memorial unto all generations. (Exod. 3:14–15)

Thus, through these revelations from God to Moses, the most sacred book in Judaism, the Torah, was given to Moses as a guide for mankind. The revealed scriptures were compiled later into five books of the Hebrew Bible—Genesis, Exodus, Leviticus, Numbers, and Deuteronomy. These books contain Jewish history and stories of man's relationship with God. The central theme of the Torah is God's commandments—a set of basic principles and laws governing human society. Of these, the following Ten Commandments are the most well-known:

> I *am* the Lord thy God, which have brought thee out of the land of Egypt, out of the house of bondage.

Thou shalt have no other gods before me.

Thou shalt not make unto thee any graven image, or any likeness of *any kind* that *is* in heaven above, or that is in the earth beneath, or that *is* in the water under the earth:

Thou shalt not bow down thyself to them, nor serve them: For I the Lord thy God *am* a jealous God, visiting the iniquity of the fathers upon the children unto the third and fourth *generation* of them that hate me;

And showing mercy unto thousands of them that love me, and keep my commandments.

Thou shalt not take the name of the Lord thy God in vain: for the Lord will not hold him guiltless that taketh his name in vain.

Remember the Sabbath-day to keep it holy.

Six days shalt thou labour, and do all thy work:

But the seventh day *is* the Sabbath of the Lord thy God: *in it* thou shalt not do any work, thou, nor thy son, nor thy daughter, thy man-servant, nor thy maid-servant, nor thy cattle, nor thy stranger that *is* within thy gates:

For *in* six days the Lord made heaven and earth, the sea and all that in them *is*, and rested on the seventh day: wherefore the Lord blessed the Sabbath-day and hallowed it.

Honor thy father and thy mother; that thy days may be long upon the land which the Lord thy God giveth thee.

Thou shalt not kill.

Thou shalt not commit adultery.

Thou shalt not steal.

Thou shalt not bear false witness against thy neighbor.

Thou shalt not covet thy neighbour's house, thou shalt not covet thy neighbor's wife, nor his man-servant, nor his maid-servant, nor his ox, nor his ass, nor any thing that *is* thy neighbour's. (Exod. 20:2-17)

Scriptures: The sacred scriptures of Judaism are called the Hebrew Bible and in Christian designation, the Old Testament. They were compiled between the twelfth century BCE and the beginning of the Common Era. The books of the Hebrew Bible are organized into three sections: the Law (Torah), the Prophets (Nebhiim), and the Writings (Kethubhim).

In addition to the Hebrew Bible, a compendium of Jewish law, called the Talmud, is also considered part of the Jewish scriptures. It contains rabbinical commentaries and the ongoing interpretation of the laws contained in the Torah.

Talmud is a guide for Jews in modeling their lives in accordance with God's laws derived from the Hebrew scriptures and their interpretation.

God of Judaism: God revealed himself to Moses by speaking to him from behind the burning bush on Mount Sinai. When asked by Moses to reveal his name, God said: "I AM THAT I AM." (Exod. 3:14). In subsequent revelations, however, God is identified through a number of names, revealing his character and relationship with mankind.

In the opening statement of the Old Testament, "In the beginning God ...," (Gen. 1: 1), the Hebrew word from which the word *God* is translated is *Elohim*. Other Hebrew name of God is Jehovah (also known as Yahweh, Yahve, or Yahveh). This name of God is translated as *Lord* in the Old Testament. Although Elohim and Jehovah are the most frequently occurring names of God in the Hebrew Bible, there are many other names and compound names that also characterize God and his attributes in the Bible.

Whereas the many names of God characterize his power and attributes, it is the basic concept of the supreme deity that characterizes a particular religion. In Judaism, this concept assigns all powers to God, infinite and without limits. For example:

1. God is the creator of the universe:
    In the beginning God created the heavens and the earth.
    And the earth was without form, and void; and darkness
    was upon the face of the deep. And the Spirit of God moved upon the face of the waters. And God said, Let there be light: and there was light. (Gen. 1:1–3)

2. God is the creator of all living things:
    So God created man in his own image, in the image of God created he him; male and female created he them.
    And God said, Behold, I have given you every herb bearing seed, which is upon the face of all the earth, and every tree in which is the fruit of a tree yielding seed; to you it shall be for meat.
    And to every beast of the earth, and to every fowl of the air, and to every thing that creepeth upon the earth, wherein there is life, I have every green herb for meat: and it was so.
    And God saw everything that he had made, and behold, it was very good. And the evening and the morning were the sixth day. (Gen. 1:27, 29–31)

3. God is one:
    Hear, O Israel: the Lord our God is one Lord. (Deut. 6:4)

Thus saith the Lord the King of Israel, and his redeemer the Lord of hosts; I am the first, and I am the last; and beside me there is no God. (Isa. 44:6)

4. God is all-powerful:
    He ruleth by his power for ever; his eyes behold the nations: let not the rebellious exalt themselves. Selah. (Ps. 66:7)

5. God is all knowing.
    God understandeth the way thereof, and he knoweth the place thereof. For he looketh to the ends of the earth, and seeth under the whole heaven. (Job 28:23–24)

Christianity

Christianity is based on the teachings of a Jewish teacher named Jesus. He was born in Bethlehem (Palestine) about two thousand years ago.[1] His birth was a supernatural event. He was born of a virgin woman named Mary, who conceived him through the power of God—the Holy Spirit. Hence, Christians believe that Jesus Christ (Christ is a title, meaning "the anointed one") had a divine father (God) and an earthly mother (Mary). In the Christian Bible, the New Testament, Jesus Christ is referred to as the Son of God.

Accounts of Jesus's life and teachings are found in the four Gospels of the New Testament: the Gospels according to Matthew, Mark, Luke, and John. These books were written about thirty-five years after the death of Jesus Christ and were based on the oral accounts of his disciples. Christians believe that the Gospels were inspired by God, and, therefore, they represent authoritative and inviolable accounts of Jesus's life and his teachings.

Jesus began his ministry when he was about thirty years old. He was a charismatic teacher, and his message struck a chord with people. He showed love and compassion to all people, especially the poor, the meek, the persecuted, and those rejected by society. In his famous Sermon on the Mount, he preached:

Blessed are the poor in spirit: for theirs is the kingdom of heaven.
Blessed are they that mourn: for they shall be comforted.
Blessed are the meek: for they shall inherit the earth.
Blessed are they which do hunger and thirst after righteousness: for they shall be filled.
Blessed are the merciful: for they shall obtain mercy.
Blessed are the pure in heart: for they shall see God.

Blessed are the peacemakers: for they shall be called the children of God. (Matt. 5:3–9)

Although Jesus's followers founded Christianity based on his teachings, Jesus did not reject the teachings of the Hebrew Bible. He said he had not come to alter the scriptures, but to fulfill them:

Think not that I am come to destroy the law, or the prophets: I am not come to destroy, but to fulfil. (Matt. 5:17)

Therefore, Christians accept the authority of the Old Testament but believe that it has been superseded by the New Testament. For example, Jesus says:

Ye have heard that it hath been said, An eye for an eye, and a tooth for a tooth: But I say unto you, That ye resist not evil: but whosoever shall smite thee on thy right cheek, turn to him the other also. (Matt. 5:38, 39)
Ye have heard that it hath been said, Thou shalt love thy neighbour, and hate thine enemy:
But I say unto you, Love your enemies, bless them that curse you, do good to them that hate you, and pray for them which despitefully use you, and perse-cute you;
That ye may be the children of your Father which is in heaven: for he maketh his sun to rise on the evil and the good, and sendeth rain on the just and on the unjust. (Matt. 5:43–45)

From the above verses and many others in the New Testament, as well as from how he lived his life, one gets the crux of Jesus's teachings: do not discard what was given to Moses and other prophets, but supplement the message with love and compassion. The love of God and the love of people, he said, are the key to one's salvation.

Jesus's ministry lasted only three years and ended with his brutal crucifixion at the hands of the Roman governor, Pontius Pilate. He was nailed through his hands and feet to a cross and left to die. After some hours on the cross, Jesus died in agony with these words:

Father, into thy hands I commend my spirit. (Luke 23:46)
Three days after his crucifixion, the Gospels say Jesus rose from the dead and told his disciples to spread his message into the world.

God of Christianity: The Bible is emphatic about the unity of God. Believing in more than one God is among the greatest of sins. Take, for example, the following passage, which is found in the Old Testament:

> Thus saith the LORD the king of Israel,
> and his redeemer the LORD of hosts; I am the first, and I am the last; and beside me there is no God. (Isa. 44:6)

The New Testament repeats this message:

> One of the scribes came ... asked him, Which commandment is the first of all?
> And Jesus answered him, The first of all the commandments is, Hear, O Israel; the Lord our God is one Lord. (Mark 12:28–29)

Trinity: Most Christians believe in one God but that he is a *triune God*, meaning three beings in one. He exists as God the Father, God the Son, and God the Holy Spirit. The concept of triune God is also called the Trinity.

The Trinity is a complex theological concept that is not biblical in origin although its underpinnings are found in the Bible. The earliest Christians insisted on the oneness of God as had been stressed in the Bible, but they also believed in Jesus Christ as the Son of God and the presence of the power of God, the Holy Spirit. The connection between the three is implied at several places in the Bible. Take the following passage from Matthew for example:

> Go ye therefore, and teach all nations, baptizing them in the name of the Father, and of the Son, and of the Holy Ghost. (Matt. 28:19)

And Paul's Letter to the Corinthians:

> The grace of the Lord Jesus Christ, and the love of God, and the communion of the Holy Ghost, be with you all. Amen. (2 Cor. 13:14)

Although the first Christians professed faith in the God of Abraham, whom they called the Father, in Jesus Christ, whom they called the Son of God, and in the Holy Ghost, whom they called the Spirit of God, they also believed in the oneness of God. It was difficult for them to explain the existence of three divine beings as one divine entity or establish any hierarchy among them without destroying the first Hebrew commandment: "I am the Lord your God. You shall have no other gods before me." (Exod. 20: 3)

To get around some of the above theological difficulties, alternate explanations were put forward, such as considering Father, Jesus, and Holy Spirit as three modes or roles of God in relation to humanity. The argument, however, was not settled until the year 325 CE when the Council of Nicaea adopted the doctrine of one essence and three beings, which is called the Trinity.

Thus, through the formula of "three beings, one substance," the Trinity links the threeness of Father, Son, and Holy Spirit with the oneness of the divine nature. According to the Trinity doctrine, the three persons of God are on equal footing. In other words, the Father is God, the Son is God, and the Holy Spirit is God. They are not three separate Gods. The three exist as one, the same substance and the same divine nature.

Admittedly the above Trinity argument is very tortuous and probably incomprehensible to most people, Christians and non-Christians alike. However, most Christians believe in Trinity as one God who is the Father, the Son, and the Holy Spirit.

The debate about the nature of Christ and Trinity was not concluded at the Nicaea Council meeting in 325 CE, however. As a matter of fact, the debate continues even to this day. For example, some ancient Christian sects, such as the Ebionites, believed in Jesus not as a "Son of God," but rather a human being who was a prophet. Many modern groups, such as Jehovah's Witnesses, the Church of Jesus Christ of Latter-day Saints, Christian Scientists, the Unification Church, and a number of other minority sects, reject the doctrine of Trinity. The mainstream Christian belief, nonetheless, is centered on the Trinity representing three divine persons of one God.

Except for the three-in-one nature of God, the God of Christianity has the same powers and attributes as described in the Hebrew Bible for the God of Abraham. Moreover, Christians believe that through the three beings of Trinity, God has revealed other aspects, such as his love for humanity. The verses in the Bible attest to God's intimate involvement with humanity:

> For God so loved the world, that
> he gave his only begotton Son,
> that whosoever believeth
> in him should not perish, but
> have everlasting life. (John 3:16)

Similarly, the Holy Spirit is sent to mankind as the comforter and teacher after the death of Christ and his ascension to God, the Father. Jesus foretells his disciples:

> But the Comforter, which is in the Holy Ghost,
> whom the Father will send in my name,
> he shall teach you all things, and
> all things to your remembrance,
> whatsoever I have said unto you. (John 14:26)

Islam

Islam, which means "submission (to God)" and by extension, "peace," arose in Arabia as a result of the teachings of the prophet Muhammad (570–632 CE). Over a billion Muslims around the world now practice it. The Muslims believe that the divine message of Islam that Muhammad received is the same message that was given to the earlier prophets, including Noah, Abraham, Moses, and Jesus. Muhammad was the last of the prophets to bring God's final message to humanity.

Muhammad was born in the Arabian town of Mecca (presently in Saudi Arabia). His father, Abdullah, died before Muhammad was born, and his mother, Amina, died when he was six years old. The orphan child was looked after by his paternal grandfather, Abdul-Muttalib, who died two years later. From then on, Muhammad came under the care of his paternal uncle, Abu Talib, who brought him up to adulthood. Abu Talib was the leader of the Hashim clan of the Quraysh tribe of Arabia.

Muhammad's genealogy has been traced back to an Arab ancestor, named Adnan, who was a descendent of Ismaeel (Ishmael), the son of Ibrahim (Abraham).

During Muhammad's time, Mecca was an important commercial center, located on one of the principal trade routes in the Middle East. It was also a religious center because of its shrine, now called Kaaba, which housed many different idols at that time. Merchants would regularly visit Mecca, combining business with the pilgrimage of Kaaba. As a result, whatever happened in Mecca could be disseminated far and wide by the trade caravans passing through Mecca.

Muhammad accompanied his uncle on trading journeys to other parts of Arabia and developed a keen interest in trade as well as in the people he interacted with. Because of his personal honesty, integrity, and truthfulness in dealing with people, he was called "the trustworthy one."

The reputation of Muhammad's honesty and trustworthiness impressed a wealthy Meccan woman, named Khadija. She hired him to run her deceased husband's business. Muhammad and Khadija married when he was twenty-five and she was forty (ca. 595 CE). There is a controversy about the number of children Khadija bore Muhammad, but his descendants trace their lineage to his daughter from Khadija, named Fatima, who is the most revered lady in the Shia branch of Islam.

Muhammad was a reserved, thoughtful, and reflective person who regularly spent nights in a cave on Mount Hira near Mecca, thinking and meditating. It was around the year 610 CE that Muhammad, while meditating in the cave, received his first revelation. Angel Gabriel (Jibril, in Arabic) brought him the first command from God:

> Read! (or Proclaim) in the name of your Lord and
> Cherisher, Who created—
> Created man, out of a (mere) clot of congealed blood:
> Proclaim! And your Lord is Most Bountiful,—
> He Who taught (the use of) the pen,—
> Taught man that which he did not know. (Qur'an 96:1–5)

Muhammad had grave doubts about the nature of his revelation. For a while, he was perplexed and terrified. As he returned home in awe of his experience, he was reassured by his wife Khadija that God would not mislead him. She consulted with her cousin, Waraqah ibn Nawfal, who declared that Muhammad, like Moses and Jesus before him, would be the Arab's prophet. Khadija and Waraqah were the first to believe that Muhammad was a prophet. They were followed by his ten-year-old cousin, Ali, his friend Abu Bakr, and others who would become his close companions and disciples in the years to come.

The Qur'an: The book of Islam, the Qur'an, is believed by Muslims to be the original word of God. It was revealed to Muhammad in parts over a twenty-year period through the mediation of Gabriel, until Muhammad's death in 632 CE.

God calls the Qur'an a part of the "Mother of the Books" that is with God (Qur'an 13:39). He also calls it a "Glorious Qur'an, (inscribed) in a Tablet Preserved" (Qur'an 85:22). Thus, the Muslims believe that the Qur'an is preserved with God and is guarded by him against any alteration or corruption.

Muslims also believe that the author of the Qur'an is God, not Muhammad. Every word of the Qur'an (originally written in Arabic) was transmitted to Muhammad by God through the angel Gabriel. Muhammad relayed each message, verbatim, to his disciples, who recorded it. Some disciples memorized it by heart. The Qur'an was reorganized and compiled in the present book form after Muhammad's death, during 650–656 CE.

About the Qur'an and its message, God says:
This is the Book: in it is guidance sure,
without doubt, to those who fear God. (Qur'an 2:2)
It is He Who sent down to you (step by step), in truth, the Book, confirming what went before it; and He sent down the Laws (of Moses) and the Gospel (of Jesus) before this, as a guide to mankind, and he sent down the Criterion (of judgment between right and wrong). (Qur'an 3:3)
He it is Who has sent down to you the Book: in it are verses basic or fundamental (of established meaning); they are the foundation of the Book: others are allegorical. But those in whose hearts is perversity—follow the part thereof that is allegorical, seeking discard, and searching for its hidden meanings, but no one knows its hidden meanings except Allah. And those who are firmly grounded in knowledge say: 'We believe in the Book; the whole of it from our Lord:' and none will grasp the Message except men of understanding. (Qur'an 3:7)
You say: We believe in God, and the revelation given to us, and to Abraham, Ishmael, Isaac, Jacob, and the Tribes, and that given to Moses and Jesus and that given to (all) Prophets from their Lord: We make no difference between one and another of them: and we bow to God. (Qur'an 2:136)

God of Islam: Although the scriptures of all the three Abrahamic religions stress unity of God, Islam is the most emphatic about it, probably because the preceding Christian concept of Trinity had deviated from the Abrahamic concept of one God. In fact, the primary difference between Christianity and Islam is the understanding of the oneness of God. In Islam, there is only one God, the God of Abraham, Moses, and Jesus. The following verses in the Qur'an make quite clear who the God of Islam is:

There is no God but He: that is the witness of God, His angels, and those endued with knowledge, standing firm on justice. There is no God but He, the exalted in Power, the Wise. (Qur'an 3:18)
God is He, than Whom there is no other God; the Sovereign, the Holy One, The Source of Peace (and Perfection), the Guardian of Faith, the Preserver of Safety, the Exalted in Might, the Irresistible, the Supreme: Glory to God! (High is He) above the partners they attribute to Him.
He is God, the Creator, the Evolver, the Bestower of Forms (or Color). To

Him belong the most Beautiful Names: Whatever is in the heavens and on earth, declares His Praises and Glory: and He is the Exalted in Might, the Wise. (Qur'an 59:23–24)

Your Guardian—Lord is God, Who created the heavens and the earth in six days, and is firmly established on the Throne (of authority): He draws the night a veil over the day, each seeking the other in rapid succession: He created the sun, the moon, and the stars, (all) governed by laws under His Command. Is it not His to create and to govern? Blessed be God, the Cherisher and Sustainer of the Worlds! (Qur'an 7:54)

He has made everything which He has created Most Good: He began the creation of man with (nothing more than) clay. (Qur'an 32:7)

But He fashioned him in due proportion, and breathed into him something of His Spirit. And He gave you (the faculties of) hearing and sight and feeling (and understanding): little thanks you give! (Qur'an 3:29)

From the above verses and others in the Qur'an, it is clear that God has the same identity and attributes as are described in the Old Testament. He also has the same attributes as God, the Father, in the New Testament. However, the God of Islam is not a triune God. Islam rejects the Trinity because it conflicts with Islam's concept of absolute oneness of God. It accepts Jesus Christ as the prophet of God but not his son. The Qur'an is emphatically clear about this:

Say: He is God, the One and Only;
God, the Eternal, Absolute;
He begets not, nor is He begotten;
And there is none like unto Him. (Qur'an 112:1–4)

They say: '(God) Most Gracious has begotten a son!'
Indeed you have put forth a thing most monstrous! At it the skies are ready to burst, the earth to split asunder, and the mountains to fall down in utter ruin.
That they should invoke a son for (God) Most Gracious.
For it is not consonant with the majesty of (God)
Most Gracious that he should beget a son. (Qur'an 19:88–92)

O People of the Book! Commit no excesses in your religion: nor say of God anything but the truth. Christ Jesus the son of Mary was (no more than) a Messenger of God, and His Word, which He bestowed on Mary, and a Spirit preceding from Him: so believe in God and His Messengers. Do not say "Trinity:" desist: it will be better for you: for God is One God: glory be to Him: (For Exalted is He) above having a son. To Him belong all things in the Heavens and on Earth. And enough is God as a Disposer of affairs. (Qur'an 4:171)

Abrahamic Faiths: How They Differ

The major difference between Judaism, Christianity, and Islam is the Trinity. Judaism regards Jesus as a teacher but not a Messiah or the Son of God. Islam believes in both Moses and Jesus as prophets. Islam also believes that Jesus was born of the Virgin Mary through the Spirit of God. Neither Christianity nor Judaism accepts Muhammad as a prophet of God. These differences appear irreconcilable. However, there is a general agreement, at least historically as told in the scriptures, that Abraham was the patriarch of the three faiths and that the God of Abraham is the God of Judaism, Christianity, and Islam.

There are many other differences and similarities between the three faiths. Table 1 summarizes these comparisons. This table was compiled by Beliefnet and ABC.com (http://www.beliefnet.com/features/abrahamicfaiths.html) and is reproduced here to help the reader understand similarities and differences side by side between the three faiths. For more in depth study of the comparisons, the reader is referred to the literature on this topic, which exists in abundance.

Table 1
The Abrahamic Faiths: A Comparison [*]

	Judaism	*Christianity*	*Islam*
God	One God	One God	One God
Central Prophet	Moses	Jesus Christ	Muhammad
Scripture	Torah, Prophets, Writings and the Talmud (oral tradition and commentary)	Bible (Old and New Testaments)	Qur'an (God's revelation to Muhammad) and Hadith (Muhammad's sayings)
War	Peace is always preferable, but war in self-defense is considered obligatory	"Just war" can be fought as a last resort; tradition of non-violent resistance	War should be fought only in self-defense and within strict limits

Table 1
The Abrahamic Faiths: A Comparison (Continued)[*]

	Judaism	*Christianity*	*Islam*
Divisions	Modern movements include Reform, Conservative, Modern Orthodox and Reconstructionist	Many theological divisions and schism: Roman Catholic, Eastern Orthodox, numerous Protestant churches	Sunni-Shiite schism based on disagreement over Muhammad's successors; broad debate over Islam's role in modern society; little theological debate
Fundamentalism	Ultra-Orthodox Jews reject the secular world and live in strict communities	Debate over literal meaning of the Bible; efforts to bring religion into daily life	Return to "pure" Islam; rejection of secular culture; efforts to bring religion into daily life and create an Islamic state
Holy City	Jerusalem	Jerusalem	Mecca
Jesus	A historic figure; not the Messiah	The Son of God	Highly respected as the second-last prophet before Muhammad
Hierarchy	No hierarchy; rabbis are considered teachers	Catholics and Orthodox have extensive hierarchy; some Protestant branches have almost none	No hierarchy; prayers are led by imams (teachers) who have studied the Qur'an
Idols and Images	Images and statues forbidden	Images and statues allowed in some denominations, but not worshipped	Images and statues forbidden
Charity	Tzedakah: 10 percent of income	Tithe: 10 percent of income	Zakat: 2.5 percent of total wealth each year

Table 1
The Abrahamic Faiths: A Comparison (Continued)[*]

	Judaism	Christianity	Islam
Proselytizing	No proselytizing; Jews must turn away would-be converts three times to insure their commitment	Conversion considered important in most traditions; Catholic and Protestant churches have missionaries	Da'wa: Muslims should share their knowledge of Islam without trying to convert. Only God can bring someone to Islam
Women	Men and women are equal in the eyes of God; traditional Judaism prescribes different roles for men and women. Orthodox men and women worship separately	Men and women worship together. Some Protestant churches ordain women as priests;	Men and women are generally treated the same in the Qur'an although women are oppressed in many Muslim cultures today. Men and women worship separately
House of Worship	Synagogue	Church	Mosque
Main Day of Worship	Saturday	Sunday	Friday
Diet	Must keep "kosher": no pork or certain seafood; other meat to be killed by kosher method; separation of meat and dairy	No dietary restrictions	No pork; other meat should be prepared by the halal method. No alcohol.
Life After Death	No immediate life after death; life in the "world to come" after the coming of the Messiah	Day of Judgment, followed by Heaven or Hell	Day of Judgment, followed by Heaven or Hell
Mysticism	Kabbalah	Numerous mystical traditions	Sufism

[*]   From: http://www.beliefnet.com/features/abrahamicfaiths.html.

PART 4

INTELLECTUAL CHALLENGES

CHAPTER 12
FAITH VS. REASON

After discussing scientific theories on the origin of the universe and human evolution, I explored the religious views on creation of the world and the concepts of Ultimate Reality and the God of Abraham in the earlier chapters. At this point, we can conclude that whereas science relies on empirical observations, experimentation, and deductions using the scientific method, religion bases its beliefs on human intuition and revelations from supernatural sources. Although the scientific and religious methods of inquiry are fundamentally incompatible, recent advances in cosmology have raised interesting questions about the origin of the universe that fall within the preview of both science and religion. In other words, science and religion are now facing each other in confrontation over who or what created the universe.

We are hearing about this conflict between science and religion much more frequently than we used to. Proponents of evolution and the believers of creationism or intelligent design are often seen at loggerheads in schools and in the public. The matter has even been brought to the courts a number of times. So at this stage of my search for truth, I am ready to address the ultimate question: Does God exist?

Does God Exist?

In order to debate the question of God's existence, I have created a candid dialogue between two hypothetical characters: person A—a believer of a personal God with all his attributes and person B—an intellectual who is guided by intellect rather than blind faith when it comes to the question of God. As a disclaimer, let me say that I am not trying to sell one idea over the other, nor am I attempting to prove or disprove God's existence. The dialogue is candid but with no ax to grind on my part. So, let us settle down and hear what they have to say.

A: God has existed in the minds of people from the dawn of civilization. Even now, billions of people around the world believe in a personal God who is the creator and sustainer of the universe. Not all these people and the civilizations before them could be wrong. From the history of the world religions, we learn that people in all cultures and societies have believed in a supreme god or gods to whom they related their life, death, and fate after death. From countless anthropological observations, one cannot avoid the conclusion that human beings are born with an innate ability or attribute that enables them to connect with their creator. Some would call this faculty the soul of man while others believe it to be the Spirit of God that links man with God from the very beginning.

B: I have heard this argument many times before—it is called the *common consent argument*. The weakness of this argument becomes apparent when we learn that belief in God is not held universally in all cultures. For example, there is no concept of God in Theravada Buddhism. Also, there are millions of people in this world who do not believe in God. As a matter of fact, even in our own culture, everyone does not believe in God. Religion in all societies is mostly learned through upbringing from families, friends, and communities. For children there is rarely a free choice in matters of adopting or rejecting a religion. Some have abandoned religion later in life when they were able to use reason rather than their inherited beliefs.

A: It is a universal law of nature that nothing happens without a cause. Every motion has a mover. Then, how could the universe have come into being without an external force or agent causing it to happen? The only explanation is that it was some great power, a great force, or a supreme being who created the universe. We call him God.

B: This argument also has a name—the *cosmological argument*. It is a weak argument because it leads to the next unanswerable question: Who created God?

Whereas, religious believers have no explanation for a God without cause, scientists do have a theory, based on established laws of physics that explains how our universe came into existence. It happened by itself! It was a spontaneous creation out of nothing and without a cause [chapter 5].

A: Life is so complex that it could not have been created by a random chance. Let us consider an analogy between the universe and a machine. If we found a watch on an uninhabited planet (and assuming that we had never seen a watch before), we would reach the unavoidable conclusion that such a complex and well-ordered object could not have been created by a random chance. It must be the work of an *intelligent designer*. The universe and life on Earth are too complex and ordered and purposeful to have been created by a random chance. One is thus forced to conclude that there must be a creator of immense power and intelligence behind all this creation. We call him the God Almighty.

B: This is a favorite argument from believers of intelligent design. Just because the universe and life on Earth are too complex for our minds to fully comprehend, there is no proof that they have been designed by an intelligent being. Fourteen billion years of cosmic and biological evolution are the true designers of the universe. From fossil evidence, a rational theory has been constructed that shows how life began with the interaction of atoms and molecules on the early earth. Complexity emerged from very simple beginnings, indeed. Molecular evolution preceded the biological evolution, which, in turn, gave rise to more complex organisms and species.

Darwin's theory of evolution and natural selection explains the species' evolution. All this is based on scientific data rather than supernatural beliefs. Life is like a giant puzzle, and the theory of evolution has provided enough scientific evidence to provide a glimpse of the mystery of its origin and development. Not all the pieces of the puzzle are in yet, but the face of life is readily discernable [chapter 8].

The theory of intelligent design, on the other hand, has no verifiable scientific data to back it up. It is purely a belief. It has been called a pseudoscience because it pretends to use scientific reasoning to prove its thesis. But even a cursory analysis of its arguments would show that it does not satisfy the scientific criteria, namely, the scientific method. It is, in fact, a creationists' religious belief masquerading itself as science. Excuse me for being so long-winded on this issue. I can't help getting a little worked up about it.

To point out a few more holes in the so-called theory of ID, let us look at the product of ID. The Grand Designer does not seem to have created a perfect world. How about the disease and natural calamities that routinely pervade life on Earth? Don't tell me they are the Designer's instruments of punishment. How about the existence of evil? Should we believe in Satan in order to explain that?

Coming back to the other assertions, namely, that the universe and life on Earth are too orderly and purposeful to have been created by chance: Well, we haven't seen the entire universe. We don't know enough about it to say whether it is orderly or purposeful. Life on Earth seems to be orderly in the sense that it follows the laws of physics, chemistry, and biology without exception. Various species of animals seem to be well-ordered, living and reproducing as if they were designed that way. But that is not proof of an intelligent designer. It is more a proof of natural selection in favor of survival of the fittest and the adaptable. Nonetheless, what is the Designer's purpose behind abnormal births: babies born with missing limbs, conjoined bodies, or no life at all? If each human being is conceived with his soul, as the creationists believe, then what about miscarriages?

Is the soul sometimes created but not given earthly life? What is the Designer's purpose behind creating a soul and not letting it live?

A: History has recorded many *miracles* that cannot be explained without the existence of God. These miracles were witnessed by people and recorded in the scriptures. How do you explain these events?

B: Almost every religion claims to have had some miracles in its history, but look at the times when these miracles happened. Were the events accurately reported or scientifically investigated to prove that they were miracles? If a supernatural event occurs today, modern science has the tools to prove it or disprove it. But none of these capabilities existed at the times when these miracles occurred. Most importantly, the believers of these miracles won't let anyone thwart their faith through investigation. Accordingly, there is no miracle in history that has been verified by scientific method and recognized as such. Miracles, like God, are based on faith, not on scientific proof.

A: Many people in the past, as well as the present, have testified that they have *experienced God*. Prophets had revelations, mystics had visions, and deeply religious people had experiences of God. Haven't you seen people closing their eyes, looking up to heaven, and gesturing as if they were experiencing God? Many say they felt as if the Holy Spirit had come into them. Why would somebody lie about such a thing?

B: The problem with revelation, mysticism, or direct experience with God is that no one else can verify these claims. Psychological analysis of some of these people has revealed that it is possible for a person to think about an object or an idea so intensely that it could produce a trancelike condition. A condition of great mental concentration can be induced by meditation, religious fervor, or mysticism, resulting in a state of altered consciousness. There is no external reality to cause this condition. It is a kind of self-hypnosis. In any case, such an experience with God is as mythical as God himself.

A: Self-awareness or *consciousness* is so unique, mysterious, and complex that it could not be explained in any physical way. God created man from elements and then gave him consciousness of himself as an individual. This consciousness is the *soul* of man that will survive even if his body perishes.

B: Consciousness is the product of brain activity [chapter 3]. Altering and destroying a part of the brain can result in complete loss of consciousness. Brain damage is known to cause a "vegetative state" in which a person can lose part or all of his or her consciousness. Where has the soul gone then? There is no logic in assuming that a supernatural power, like God, is controlling one's consciousness and not the brain itself.

A: It is a safe bet to believe in God whether he exists or not. If you believe and he does not exist, there is no harm done. But if you do not believe and he does exist, you are surely going to hell.

B: This is a funny argument, known as *Pascal's Wager*. The problem with Pascal's Wager is that it does not offer any proof of God's existence. All it does is play upon your fear of hell. The argument is devoid of any intellectual honesty or integrity.

Do We Need God?

A: God created man in his own image. He gave man *morality* so that he would know the difference between right and wrong. He gave man free will so that he could choose between good and evil by his own volition. The scriptures are testament to the absolute divine moral code that has guided humanity since its creation. There is nothing in the evolution theory that explains how humans have acquired moral principles of right and wrong. The Holy Scriptures provide the evidence that morality came from God through revelations that were sent to mankind from the time of Adam and Eve to present day.

B: All religious scriptures contain moral codes for their believers. Prophets and reformers have used God's authority to bring about social change and have introduced many "God-given" moral codes to eliminate the social ills of society. For example, the message of love, compassion, and social justice is found in most of the scriptures. There is no doubt that such principles, with the backing of God's authority, have played a positive role in nurturing kindness, justice, and social harmony in various human societies. However, one can also postulate that humans are perfectly capable of creating such morality on their own. Human minds are quite creative, and they are endowed with emotions that are the building blocks of human morality. Through the ages, humans have strived to develop rules of social interaction and conduct in order to progress socially, economically, and politically. Many of their laws have incorporated moral codes from the scriptures as well as from their imaginative, artistic, and intellectual inventiveness. Because of the fact that there are thousands of scriptural and nonscriptural writings with thousands of moral codes, moral sensibility seems to be a fundamental trait of human nature.

Whereas scriptures are considered to be the fountainheads of religious morality, they are by no means perfect. Inconsistencies and contradictions have been found in every scripture, although the faithful either deny them, ignore them, or get around them with favorable explanations. Factual inconsistencies aside, some of God's actions, as described in the scriptures, seem to be at odds with the accepted standards of morality in the modern world. Let me point out a few as examples:

God's love for humanity: Granted, the Bible says:
For God so loved the world that He gave His only begotten Son, that whosoever believeth in him should not perish, but have everlasting life. (John 3:16)
Granted, the opening chapter of the Qur'an starts with the verses:
In the name of Allah, Most Gracious, Most Merciful.
Praise be to Allah, the Cherisher and Sustainer of the Worlds.
Most Gracious, Most Merciful. (Qur'an 1:1–3)

Granted, there are numerous verses in the Old Testament, the New Testament, and the Qur'an confirming that God loves humanity and that he is the most loving, the most forgiving, and the most merciful.

But, some actions of God, as reported in the Bible as well as in the Qur'an, run counter to our current moral sensibilities. For example, how could the most merciful God destroy the cities of Sodom and Gomorrah with "brimstone and fire from the LORD out of heaven" (Gen. 19:24–26).

Or, if "God so loved the world" (John 3:16), how could he drown the whole planet and destroy everything with a flood (Noah's flood) as told in the Bible?

And every living substance was destroyed which was upon the face of the ground, both man and cattle, and the creeping things, and the fowl of the heaven. (Gen. 7:23)

And, what about God's commandment:

Whosoever doeth any work in the Sabbath-day he shall surely be put to death. (Exod. 31:15)

God's punishment for nonbelievers or sinners is vividly described in several places in the Bible as well as in the Qur'an. According to one description of hell:

Truly Hell is as a place of ambush,
For the transgressors a place of destination:
They will dwell therein for ages.
Nothing cool shall they taste therein, nor any drink,
Save a boiling fluid and a fluid, dark,

Murky, intensely cold,
A fitting recompense (for them). (Qur'an 78:21–26)

From these verses and many others like them in the Holy Scriptures, it becomes apparent that God's love, mercy, and kindness are reserved mostly for the faithful. For the nonbelievers or the sinful, however, there is nothing but the wrath of God with all its rage and fury. The description of hell for the sinners exceeds the human concept of punishment for any crime, no matter how grievous. Hell is a torture by God for eternity, for which there is no appeal to a higher authority or another supreme court of justice.

Slavery: The buying and selling of fellow human beings into bondage and servitude is one of the most immoral practices in human history. Thanks to the enlightenment movement of the last century or so, the world is almost unanimous, at least overtly, in its condemnation of slavery. Any form of slavery is considered repugnant to the modern standards of morality. This is not so in the Bible. The Qur'an actively discourages the practice but does not ban it outright. It is simply mind-boggling that God, the Creator of all beings, did not forbid the inhuman practice of slavery from the very beginning. He, in fact, condoned it in the Old Testament as well as the New Testament. Let us look at a few samples:

> If a man smite his servant, or his maid with a rod, and he die under his hand; he shall be surely punished. Not withstanding, if he continue a day or two, he shall not be punished: for he is his money. (Exod. 21:20–21)
> Then you shall take an awl, and thrust it through his ear unto the door, and he shall be thy servant forever. And also unto thy maidservant you shalt do likewise. (Deut. 15:17)
> Jesus was asked to visit and heal a slave who was sick. Jesus did not say anything to free the slave. Instead he complimented the slave owner for his faith: I say unto you, I have not found so great faith, no, not in Israel. (Luke 7:9)
> Regarding the duties of slaves, the Bible says:
> Servants, be obedient to them that are your masters according to the flesh, with fear and trembling, in singleness of your heart, as unto Christ. (Eph. 6:5)

The Qur'an did not ban slavery outright as it banned pig's meat, alcohol, idolatry, and many other ungodly customs and rituals. However, it did promote the freeing of slaves. Here is a sample:

> It is not righteousness that you turn your faces toward East or West, but it is righteousness ... to spend of your substance ... for the ransom of slaves. (Qur'an 2:177)

Verily we have created Man into toil and struggle ... And what will explain to you the path that is steep? (It is) freeing the bondman. (Qur'an 90:4–13)

Status of women: Women have been socially subjugated in most societies and religions have a lot to do with their suppression. Most major religions consider men to be superior to women, and some have condemned women in the strongest terms. Also, if there were any rights given to women in the scriptures, they were effectively taken away by social laws and customs. Religions, being mostly male dominated, have contributed to the suppression of women by making them socially and economically dependent on their husbands. This was all done by the authority of God. I will present a few samples of discriminatory verses that occur in the scriptures of five major religions: Hinduism, Buddhism, Judaism, Christianity, and Islam.

These are examples found in Hindu scriptures:
In childhood, a female must be subject to her father, in youth to her husband, when her lord is dead, to her sons; a woman must never be independent. (Laws of Manu 5:148)[1]
Even if high-born and gifted with beauty and possessed of protectors, women wish to transgress the restraints assigned to them. This fault truly attaches to them, O Narada. There is nothing more sinful than women. Verily the women are roots of all evils. (Mahabharata, ed. Dutt, 13:38:11–13)[2]
This is an example found in Buddhism:
You should know that when men have close relationships with women they have close relationships with evil ways....
Fools lust for women like dogs in heat....
Women can ruin the precepts of purity.
They can also ignore honour and virtue.
Causing one to hell, they prevent rebirth in heaven.
Why should the wise delight in them? ...
Ornaments on women show off their beauty.
But within them there is great evil as in the body there is air....
The dead snake and dog are detestable,
But women are even more detestable than they are....
Confused by women, one is burnt by passion.
Because of them one falls into evil ways.
There is no refuge. (Speech of Buddha to King Udayana, From the Maharatnakuta)[2]
The Old Testament says:

And the daughter of any priest, if she profane herself by playing the whore, she profaneth her father: she shall be burnt with fire. (Lev. 21:9)

When men strive together one with another, and the wife of one draweth near for to deliver her husband out of the hand of him that smiteth him, and putteth forth her hand, and taketh him by the secrets [genitals]: then thou shalt cut off her hand, thine eye shall not pity her. (Deut. 25:11–12)

In the New Testament, Jesus says:

Behold, I will cast her into a bed, and them that commit adultery with her into great tribulation, except they repent of their deeds. And I will kill her children with death; and all the churches shall know that I am he which searcheth the reigns and hearts: and I will give unto every one of you according to your works. (Rev. 2:22–23)

And, in the words of Paul:

Wives, submit yourselves unto your husbands, as unto the Lord. For the husband is the head of the wife, even as Christ is the head of the church: and he is the saviour of the body. Therefore as the church is subject unto Christ, so let the wives be to their husbands in every thing. (Eph. 5:22–24)

In Islam, according to the Qur'an, men are head of the household. They have authority over women:

Men are the protectors and maintainers of women, because Allah has given the one more (strength) than the other, and because they support them from their means. Therefore the righteous women are devoutly obedient, and guard in (the husband's) absence what Allah would have them guard. As to those women on whose part you fear disloyalty and ill-conduct, admonish them (first), (next), refuse to share their beds, (and last), beat them (lightly); but if they return to obedience, do not seek against them means (of annoyance): for Allah is Most High, Great (above you all).

If you fear a breach between them twain, appoint two arbiters, one from his family, and the other from hers; if they wish for peace, Allah will cause their reconciliation: for Allah has full knowledge, and is acquainted with all things. (Qur'an 4:34–35)

A divorce is only permissible twice: after that the parties should either hold together on equitable terms, or separate with kindness. It is unlawful for you, (men), to take back any of your gifts (from your wives), except when both parties fear they would be unable to keep the limits ordained by Allah. (Qur'an 2:229)

The Qur'an allows men to have four wives, but it stresses equal treatment:

> If you fear that you shall not be able to deal justly (with them), then only one ... (Qur'an 4:3)

Thus, a careful reading of the scriptures would indicate that in some matters the scripture-based morality is clearly at odds with modern standards of morality. If God's morality is an absolute standard, then how do we explain the wanton destruction of cities and of the whole planet (by drowning) just because the inhabitants sinned? How could God, the fountain of mercy, reserve an awful place like hell for the nonbelievers? Isn't it odd that there is not a single commandment banning slavery? Are all men and women created equal or is it just men?

Finally, to answer the question: do we need God because we need morality? One should ask the faithful to explain the moral glitches of the "divine scriptures" before naming God as the source of human morality. Or they should concede that the scriptures and the moral commandments they contain are the works of humans, constrained by existing cultures and norms, rather than any divine inspiration. Or, they should accept the view that the moral codes of the scriptures should be amended from time to time to meet the needs of the evolving human societies.

CHAPTER 13
ALTERNATIVES TO GOD

The idea of multiple gods or one personal God did not receive universal acceptance, either in the past civilizations or in the present one. There have always been skeptics who questioned the logic of supernatural phenomena. The reason for skepticism is the rationality of the human mind.

Human beings have the ability to think rationally and logically. They also have a "heart" which symbolizes their emotions such as love, hate, anger, and fear. In accepting or rejecting a supernatural idea like that of gods or a God, both the rational and emotional attributes come into play. If you are guided by scientific rationality, you reject anything that is supernatural because it has no physical reality that can be verified. On the other hand, if you are persuaded by your emotions, you accept God as a matter of faith. In matters of supernatural beliefs, faith and reason are diametrically opposed. Therefore, depending on which one has the upper hand—your faith or your scientific reasoning—you are a believer or a nonbeliever, accordingly.

The nonbelievers have been called by many names: infidels, agnostics, atheists, humanists, secular humanists, and a variety of other designations, all implying their denial of God to a varying degree of emphasis or stigma. Many of these nonbelievers stay in the closet—they do not discuss their beliefs in the open. Others are more open, while some have organized themselves into social movements or ideologies with specific ethical philosophies. They represent alternatives to the faith-based religions.

Agnosticism

Thomas Huxley coined the term *agnosticism* in 1869 to describe those who do not believe in the existence or nonexistence of deities outside the realm of natural phenomena. An agnostic maintains that the human mind is incapable of knowing the truth about God. Huxley describes his view on agnosticism as follows:

> I neither affirm nor deny the immortality of man. I see no reason for believing it, but, on the other hand, I have no means of disapproving it. I have no *a priori* objections to the doctrine.... So I took thought, and invented what I conceived to be the appropriate title of 'agnostic.' It came to my head as suggestively antithetic to the 'gnostic' of Church history, who professed to know so much about the very things of which I was ignorant. To my great satisfaction the term took.[1]

Huxley maintained that the truth about God and religion are unknowable, and therefore he professed noncommittal about either the belief or nonbelief. However, he strongly believed in honest pursuit of truth:

> In matters of the intellect, follow your reason as far as it will take you, without regard to any other consideration. And negatively: In matters of the intellect, do not pretend that conclusions are certain which are not demonstrated or demonstrable.[2]

There is a subtle distinction between agnosticism and atheism. Whereas, agnosticism maintains that the truth about God (or gods) is unknowable, atheism denies the existence of any deity outright. The two views, however, are not too far apart. An agnostic, although uncommitted to the existence or nonexistence of God, is not any closer to believing in God than an atheist is. They both effectively represent the same position, namely, that there is no evidence for the existence of any supernatural being. In the common usage of the terms, however, agnosticism is considered more moderate in view than atheism. The former has doubts whether God exists or not, but the latter is a nonbeliever without further qualifications.

Atheism

Atheism denies the existence of God or gods and rejects all religious beliefs based on supernatural revelations. The atheist position is that since there is no empirical evidence for any gods, there is no need to believe in their existence in order to explain the universe or lead a moral life. In other words, *absence of evidence is evidence of absence.*

The argument of atheism has been articulated by many philosophers, scientists, intellectuals, politicians, literary figures, artists, and freethinkers. A few are presented below to help understand the logic of atheism. It is left to the reader to agree or disagree with these arguments.

> Albert Einstein, physicist (1879–1955) said:
> I do not believe in a personal God and I have never denied this but have expressed it clearly. If something is in me which can be called religion then it is the unbounded admiration for the structure of the world as far as our science can reveal it.[3]
> Dr. James Watson, biologist, discoverer of DNA remarked:
> I don't think we're here for anything, we're just products of evolution. You can say 'Gee, your life must be pretty bleak if you don't think there's a purpose' but I'm anticipating a good lunch.[4]
> Friedrich Nietzsche, philosopher (1844–1900) wrote:
> All religions bear traces of the fact that they arose during the intellectual immaturity of the human race—before it had learned the obligations to speak the truth. Not one of them makes it the duty of its God to be truthful and understandable in his communications.[5]

John Adams, U.S. president and founding father of the United States of America (1700–1836) stated:

> What influence in fact Christian ecclesiastical establishment had on civil society? In many instances they have been upholding the thrones of political tyranny. In no instances they have been seen as the guardians of the liberties of the people. Rulers who wished to subvert the public have found in the clergy convenient auxiliaries. A just government, instituted to secure and perpetuate liberty, does not need the clergy.[6]
> Frank Lloyd Wright, American architect (1869–1959) said:
> I believe in God, only I spell it Nature.[7]

Humanism

The word *humanism* ordinarily means human nature or the quality of being human. But as a social philosophy, it has a specific meaning: an ethical system that affirms the nature, dignity, and ability of man to determine his moral and ethical conduct without recourse to supernaturalism.

Historically, ethics and morality have been the main forte of world religions. Even to this day, most people identify moral values with religious virtues. However, there have been undercurrents of separating the two all along. The development of ethical morality as part of philosophy can be traced back to ancient China, classical Greece, and Rome. The Renaissance of the fourteenth century and the Enlightenment of the eighteenth century in Europe greatly revived art, literature, and philosophy. It gave impetus toward rationality and learning, free from prejudice and superstition. The scientific revolution of the nineteenth and twentieth centuries instilled new confidence in man's ability to solve social, economic, and political problems through scientific reason and logic. Humanism is the culmination of these broad scholarly movements. It is the product of social, cultural, and political enlightenment that started centuries ago and continues through modern times.

Although humanism rejects transcendental justification of morality, it endorses moral principles derived from any source, provided they are based on the commonality of human nature and interests. Humanism is nonreligious, not antireligious.

Humanism stresses critical reasoning, factual evidence, and compassion in seeking solutions to human problems. In particular, it avoids dogmas and ideologies that can cloud the issues and prevent one from seeking an objective truth. There are no absolute doctrines of humanism, but its widely accepted principles have been enunciated in the Humanist Manifesto. The following humanism precepts, extracted from the Humanist Manifesto, are briefly stated. For details, the reader is referred to the original texts.[8]

1. Knowledge of the world is derived by observation, experimentation, and rational analysis.

2. Humans are an integral part of nature, the result of unguided evolutionary change.

3. The responsibility for our lives and the kind of world in which we live is ours and ours alone. We firmly believe that humanity has the ability to progress toward its highest ideals.

4. We appreciate the need to preserve the best ethical teachings of traditional religions, but reject those features that deny humans a full appreciation of their own potentialities and responsibilities.

5. Promises of immortal salvation or fear of eternal damnation are both illusory and harmful. There is no credible evidence that life survives after the death of the body.

6. We affirm that moral values derive their source from human experience, needing no theological or ideological sanctions.

7. Critical intelligence, infused by a sense of human caring, is the best method that humanity has for resolving problems. Reason should be balanced with compassion.

8. The preciousness and dignity of the individual person is a central humanist value.

9. Short of harming others or compelling to do likewise, individuals should be permitted to express their sexual proclivities and pursue their lifestyles as they desire. The right to birth control, abortion, and divorce should be recognized.

10. Safeguarding of human freedom and human rights with a full range of civil liberties must be the goal of all societies.

11. Commitment to an open and democratic society is essential to developing goals and values for the society.

12. Separation of church and state, with maximum freedom for different moral, political, religious, and social values in a society, is imperative.

13. Economic systems should not be governed by rhetoric or ideology but by whether or not they increase economic well-being of all individuals and groups.

14. The principle of moral equality must be furthered through elimination of all discrimination based on race, religion, sex, age, or national origin. Availability of equal opportunity requires recognition of talent and merit. Society should provide basic economic, health, and cultural needs for those who are unable to provide for themselves. The right to universal education must be affirmed.

15. We should work toward eliminating division of humankind on nationalistic grounds with the ultimate goal of building a world community in which all sectors of the human family can participate.

16. This world community must renounce the resort to violence and force as a method of solving international disputes. War is obsolete and so is the use of nuclear, biological, and chemical weapons.

17. The cultivation and conservation of nature is a moral value. Exploitation of natural resources, uncurbed by social conscience, must end.

18. Extreme disproportions in wealth, income, and economic growth should be reduced on a worldwide basis. World poverty must cease.

19. We should resist any moves to censor basic scientific research on moral, political, or social grounds, provided the technology is judged to have no harmful consequences of its use.

20. We should expand communication and transportation across frontiers. We call for all international cooperation in culture, science, the arts, and technology across ideological borders.

Secular Humanism

The broad category of humanism can be divided into secular and religious types. Secular humanism, as the name suggests, is a humanist philosophy that explicitly rejects theistic morality, divine revelation, religious dogma, and the existence of any supernatural phenomenon or being. It stresses the fact that the religious view of absolute morality, ordained by God or gods, cannot be supported rationally and, therefore, could not be made the basis of human morality, laws, and ethics. Secular humanists consider themselves as the true humanists and deny that religious humanists (discussed later) qualify as genuine humanists.

Most of the humanist principles, described earlier in the Humanist Manifesto, also apply to secular humanism. Although the core beliefs of humanism and secular humanism are essentially the same, there is a subtle distinction. This has to do primarily with the degree of emphasis that the latter puts on secularism, meaning rejection of any form of religious faith or worship.

In spite of wide differences of opinion among secular humanists, there is a loose consensus on several principles that have been enunciated in a document, *A Secular Humanist Declaration*, issued in 1980 by the Council for Democratic and Secular Humanism (now the Council for Secular Humanism). The following principles are extracted from the Declaration document and are stated here only briefly to highlight the main tenets of secular humanism.[9]

1. Free inquiry: We oppose any attempt or effort by ecclesiastical, political, ideological, or social institutions to shackle free thought.

2. Separation of church and state: Because of their commitment to freedom, secular humanists believe in the principle of the separation of church and state. A pluralistic, open society allows all points of view to be heard. Any effort to impose an exclusive conception of truth, piety, virtue, or justice on the whole society is a violation of free inquiry.

3. The ideal of freedom: We defend the ideal of freedom, not only freedom of conscience and belief from those ecclesiastical, political, and economic interests that seek to repress them, but genuine political liberty, democratic decision making based on majority rule, and respect for minority rights and the rule of law.

4. Ethics based on critical intelligence: For secular humanists, ethical conduct is, or should be, judged by critical reason. Although we believe in tolerating diverse lifestyles and social manners, we do not think they are immune to criticism. Nor do we believe that any one church should impose its views of moral virtue and sin, sexual conduct, marriage, divorce, birth control, or abortion, or legislate them for the rest of society.

5. Moral education: We believe that moral development should be cultivated in our children and young adults. We do not believe that any particular sect can claim important values as their own exclusive property. Hence, it is the duty of public education to deal with these values. Accordingly, we support moral education in the schools that is designed to develop an appreciation of moral virtues, intelligence, and the building of character. Secular humanism is not so much a specific morality as it is a method for the explanation and discovery of rational moral principles.

6. Religious skepticism: Secular humanists may be agnostics, atheists, nationalists, or skeptics, but they find insufficient evidence for the claim that God has intervened miraculously in history or revealed himself to a chosen few or that he can save or redeem sinners. They believe that men and women are free and are responsible for their own destinies and that they cannot look toward some transcendent Being for salvation. We have found no convincing evidence that there is a separable "soul" or that it exists before birth or

survives death. We must therefore conclude that the ethical life can be lived without the illusions of immortality or reincarnation.

7. Reason: We are committed to the use of rational methods of inquiry, logic, and evidence in developing knowledge and testing claims to truth. Since human beings are prone to err, we are open to the modification of all principles, including those governing inquiry, believing that they may be in need of constant correction.

8. Science and technology: We believe the scientific method, though imperfect, is still the most reliable way of understanding the world. Hence, we look to the natural, biological, social, and behavioral sciences for knowledge of the universe and man's place within it. We appreciate the great benefits that science and technology (especially basic and applied research) can bring to humankind, but we also recognize the need to balance scientific and technological advances with cultural explorations in art, music, and literature.

9. Evolution: We deplore the efforts by religious fundamentalist (especially in the United States) to invade the science classrooms, requiring that creationist theory be taught to students and requiring that it be included in biology textbooks. This is a serious threat to academic freedom and to the integrity of the educational process. We do not deny the value of examining theories of creation in education courses on religion and the history of ideas, but it is sham to mask an article of religious faith as a scientific truth and to inflict that doctrine on the scientific curriculum. If successful, creationists may seriously undermine the credibility of science itself.

10. Education: In our view, education should be the essential method of building humane, free, and democratic societies. We need to embark upon a long-term program of public education and enlightenment concerning the relevance of the secular outlook to the human condition.

11. Conclusion: We believe that it is possible to bring about a move to a humane world, one based on the methods of reason and the principles of tolerance, compromise, and the negotiations of difference. We deplore the growth of intolerant sectarian creeds that foster hatred. In a world engulfed by obscurantism and irrationalism, it is vital that the ideals of the secular society not be lost.

Religious Humanism

Religious humanism, like secular humanism, has the same core principles as stated in the Humanist Manifesto, discussed earlier. The primary distinction between the religious and secular forms of humanism is that the former integrates religious rituals with the humanist philosophy while the latter divorces itself from religious faith altogether.

Religious humanism is a bit confusing in the sense that it follows the secular principles of humanism and yet attempts to fuse religious customs into scientific humanist inquiry. The name *religious humanism* is also somewhat misleading. There is nothing religious about a philosophy that believes in secular principles of morality as espoused by humanism.

Another view of religious humanism is to consider humanism as a religion, fulfilling all of the functions of a revealed religion such as rituals and moral guidance but without the belief in God or gods or divine revelations. According to this concept of religious humanism, conscience is the soul of the believer. True religion and spirituality reside within humans themselves. Through intelligence, rationality, and a sense of right and wrong, a human being has the ability to develop moral standards. This is consistent with the secular humanists' view that all religions are human creations that contribute to human morality and social well-being. Because they are human creations, these moral codes and doctrines need perpetual corrections to meet human needs as they change with time. Divine intervention in this process is not required nor is it believed to occur.

In the Humanist Manifesto I (1933),[10] the terms *humanism* and *religious humanism* are used interchangeably, although *secularism* is assumed as the basic principle of all humanist philosophies. The following numbered declarations (extracted from the Humanist Manifesto I and briefly stated here) specifically refer to religious humanism:

- EIGHTH—Religious humanism considers the complete realization of human personality to be the end of man's life and seeks its development and fulfillment in the here and now. This is the explanation of the humanist's passion.

- NINTH—In the place of the old attitudes involved in worship and prayer, the humanist finds his or her religious emotions expressed in a heightened sense of personal life and in a cooperative effort to promote social well-being.

- TENTH—It follows that there will be no uniquely religious emotions and attitudes of the kind hitherto with belief in the supernatural.

- TWELFTH—Believing that religion must work increasingly for joy in living, religious humanists aim to foster the creative in man and to encourage achievements that add to the satisfactions of life.

- THIRTEENTH—Religious humanism maintains that all associations and institutions exist for the fulfillment of human life. Certainly, religious institutions, their ritualistic forms, ecclesiastical methods, and communal activities must be reconstituted as rapidly as experience allows in order to function effectively in the modern world.

In the concluding statement, the Humanist Manifesto I declares:

So stand the theses of religious humanism. Though we consider the religious forms and ideas of our fathers no longer adequate, the quest for the good life is still the central task for mankind. Man is at last becoming aware that he alone is responsible for the realization of the world of his dreams, that he has within himself the power for its achievement. He must set intelligence and will to the task.

PART 5

AT THE CROSSROADS
OF FAITH AND REASON

CHAPTER 14

CREATED BY GOD OR BY CHANCE?

One of the key attributes of God, as claimed by many religions, is that he is the creator of the universe. The question of God's existence is therefore intimately connected with the question of who created the universe. Understanding the universe is akin to understanding the reality of God.

In my search for the reality of God, I first examined the scientific view of the universe: how it was created, how it evolved, and what lies in its future. I also looked at the equally complex questions of how life got started on Earth and how it evolved into its innumerable forms and species, including humans. We call the scientific views as theories because in science the quest for truth never ends. No matter how established a theory might become, it never reaches a point of unquestionable fact or an eternal truth.

To cite an example, Newton's theory of gravity was considered to be irrefutable for 2½ centuries until Einstein introduced his general theory of relativity, which found some faults with Newton's theory under extreme conditions and changed the whole paradigm of gravity as a Newtonian force. In the same way we expect, at least in principle, that there would be future modifications of Darwin's theory of evolution, although his theory is well established in scientific circles at the present time. However, it must be acknowledged that this fallibility of science is not a weakness. It is in fact its greatest asset.

After gaining scientific understanding of how the universe sprung into existence out of nothing, we turned our focus on the religious view of creation. From the Hindu scriptures, the Upanishads, we learned that Brahman is the Ultimate Reality of all worlds and the source of all things and all beings. In the more ancient Hindu scriptures, the *Rig-Veda*, we read about how the universe was created out of the void. From several other religions of the ancient times, we acquainted ourselves with the transcendental view of the world and of the Ultimate Reality, which is believed by some to be an impersonal force or deity that created the universe. From the great monotheistic faiths, we learned about the one God, the God of Abraham, and read the description in the Bible and in the Qur'an of how he created the universe. These accounts pointed to God as the sole creator of the universe and of life.

Let us examine more closely the question of whether God is indeed the creator of the universe or the universe was created by random chance.

Did God Create the Universe?

The story of creation as told in the scriptures is brief but dramatic, in the literary sense. The words of Genesis, for example, have the beauty, the emotion, and the poetry worthy of this great moment—the moment of creation. However, there is no physical explanation as to how the universe actually came into being or how matter and energy were created or if the universe was created in a finished form as it exists today or only as a seed that was implanted and grew into the vast universe we observe today. It seems the author of the biblical story, if it was God, did not want us to know more than the fact that he created the universe from void in six days. Maybe he wanted us to figure out the rest on our own.

No matter how much you read into the scriptural stories of creation, you don't get a clue as to how the universe physically came into existence. I have read many theological explanations and scholarly interpretations of the scriptural accounts of creation, but none gave me any insight into how the universe actually came into being. There is nothing in the words of the scriptures from which we can construct a scientific theory of creation. You just have to believe in what God said and that is the end of the story. So, the realization soon sets in that the religious scriptures are ancient myths, not cosmology theories.

That brings up an important question. Should we believe in the scientific theories of creation or be content with the religious myths? Well, if one is to believe that the religious account is the absolute truth, then we should realize that we are at a dead end. On the other hand, if we give credence to science, the religious belief may be in jeopardy. For a scientist, who may also believe in religion, it presents a great dilemma. A scientist is not supposed to have any bias in the study when searching for a truth. So how could a cosmologist research into how the universe was created if he or she also believes in the biblical account as the unalterable truth? A humorist may say: find another line of work!

Because the religious accounts are not at all helpful in understanding the universe or its creation process, we have no other recourse but to explore the mysteries of the universe through our own scientific inquiry. Cosmology is the science we turn to if we want answers, remembering very well that science, unlike religions, must be free of supernatural beliefs. Nor should there be a dead end to its inquiries. The creation of the universe by quantum fluctuation and its expansion and evolution through the big bang model are currently the most widely accepted scientific theories. The door is not shut against another challenging view or model, provided it meets the standards of the scientific method (chapter 3). The theory of intelligent design, however, does not qualify to be a legitimate challenger because it does not meet the scientific criteria. There is simply no room in science for unsubstantiated beliefs or opinions.

The scientific theories of creation and cosmic evolution were discussed in chapters 5–7. Let us briefly reiterate the evidence:

1. Spontaneous creation of the universe out of nothing (no matter, no energy, no space, and no time) by quantum fluctuation does not violate the physical laws of nature. Such phenomena, such as the Casimir effect and the Lamb-Rutherford shift phenomenon, have been observed in the laboratory settings. Although energy was created out of nothing, the laws of conservation of energy have not been violated. Astronomical data show that the positive energy associated with the total mass of the universe is equal in magnitude to the negative energy locked away in its gravitational field. The net energy of the universe is still zero, as it was before its beginning.

2. Observed expansion of the universe with time as measured by Hubble and others.

3. Measurement of cosmic microwave background radiation (CMBR), which correlates with the average temperature of the universe.

4. Relative abundance of hydrogen and helium measured at present fits the prediction of the theory.

5. Agreement exists between the model and the observed cosmic evolution of stars, galaxies, planets, etc.

6. A vast amount of astronomical data is supportive of the big bang model.

Comparing the religious stories of creation, as told in the various scriptures, with the scientific model, we are faced with the following options: (1) believe in the religious version of creation and disregard others that disagree; (2) believe in the scientific theory and disregard any other theory that does not meet the standards of scientific method; or (3) believe in both the religious and scientific views, with the presumption that the religious stories are allegorical and do not conflict with the scientific theories.

There are many scientists as well as religious people who are quite comfortable with option 3. They acknowledge that God did not reveal all the secrets of the universe to humanity and has left the door open for humans to further investigate. Many of these people also believe that the stories in the scriptures are metaphorical truths and should not be taken literally.

However, we cannot prove whether the scriptural accounts are literal or metaphorical. It is a matter of interpretation or a personal opinion, just like most things are in religions. For that reason, therefore, it is almost impossible to pin down religious facts for scientific examination. If an obvious contradiction is found (e.g., the sun moving around the stationary earth, the earth being flat, the age of the earth, and the time of first appearance of man), there is often an explanation from some believers to reconcile the scriptural accounts with the irrefutable scientific evidence. For example, some believers maintain that God's timescale is different than that of humans. And, if easy reconciliation is not possible, the believers may just call the scriptural description as metaphorical or miraculous. Jesus Christ's birth of the Virgin Mary contradicts reproductive science, but God, by definition, is not bound by physical laws. After all, he created Adam from clay without any parent.

We could go on and on pointing out contradictions between religion and science, but it will not get us any closer to reconciling religion with science. We must recognize that religion is based on faith without reason, and science is based on reason without faith. The two have different sources and methodologies to arrive at the truth, which is often not the same.

Did God Create Life?

The second key attribute of God is that he is the creator of life. In both the Hindu and Abrahamic religions, the scriptures maintain that the universe, including Earth, plants, animals, and humans, was created by a supreme being. Those who believe literally in the scriptural accounts reject the ideas of spontaneous creation of the universe and the unguided evolution of life on Earth. They believe that the act of creation was a divine intervention. There are some religious believers, however, who consider the scriptural account of creation as a figure of speech and do not dwell on contradictions between science and the literal meaning of the scriptural texts.

One of the contradictions between the literal interpretation of creation according to Genesis and the scientific view is the age of the earth. The literal believers, known as the young-earth creationists, reject the scientific view that the earth is 4.6 billion years old. They believe that God created the earth, plants, and animals in six days, about six thousand years ago. They base their geological times through biblical genealogies and the accounts of a global flood as told in the Bible.

The young-earth creationists reject the scientific theory of evolution because they believe that all land animals were destroyed in the global flood, except those that were saved on Noah's ark. Accordingly, the present animals, with all their diversity, are descendants of those that survived the flood.

The nonliteral creationists do not dispute the age of the universe or the earth as determined by science, but hold the view that the universe was created by God and that the creation accounts of Genesis are not to be taken literally. They, however, doubt the evolutionary theory and believe that the universe, as it exists now, and all life in it, was created and guided by God.

Although not all creationists hold exactly the same beliefs, they have one belief in common, namely, that it is God who created the universe and whatever is in it. He is the one who guides everything in the universe to its appointed course. He guides nature through his laws and intervenes in its workings as he deems fit.

The doctrine of creationism is common in many religions, especially in the Abrahamic faiths. According to a 2001 Gallup poll, about 72 percent of Americans believe in creationism. The same survey showed that about 45 percent of Americans believe that God created humans within the last ten thousand years. In the Western world outside the United States, however, creationism vs. evolution is not a big issue. The scientific theory of evolution is the commonly held view and taught in schools without any controversy.

The antagonism between science and religion over the issue of creation and evolution arises primarily because of their different approaches to explaining the natural world. The sciences use the scientific method (discussed in chapter 3) to discover and explain facts about the natural world. They formulate theories after systematic and objective collection of data, experimental testing, and verification. The scientific theories are open to further investigation and, if proven incorrect, are discarded or modified. A theory reigns until it fails the scientific test.

Creationism, on the other hand, starts with a conclusion, based on scriptural accounts or religious tradition, and then uses scriptural or theological argument to provide supporting evidence. In science, such logic is untenable. You cannot explain one unknown with another unknown or one mystery with another mystery.

Therefore, creationism is not a theory or hypothesis to be tested. It is a belief based on the scriptures, presumably revealed to man by God. These writings derive their authority from God, not from scientific data. The accounts of creation may vary in different scriptures, but in the minds and hearts of their respective believers, they represent absolute truths and are therefore inviolable.

In an attempt to give creationism a scientific character, a concept called intelligent design (ID) has been introduced. Its proponents call it a scientific theory and claim that it is an alternative theory to evolution regarding the origin of life. The stated objective of the ID theory is to investigate whether or not life was designed by an intelligent being.

The concept of ID was discussed in chapter 8. Let us recall its main arguments:

1. The life systems are "irreducibly complex," meaning that they are composed of well-matched interacting parts to support a biological function. Any attempts to further analyze or dissect will cause them to cease functioning. The systems must therefore have been deliberately engineered by an intelligent being. Evolution cannot account for the emergence of these incredibly complex units of biological function.

2. Living organisms embody "complex specified information," which has a negligibly small probability of occurring by chance (e.g., a probability of 1 in 10^{150}).

3. According to the anthropic principle, which supports ID, all natural laws, physical constants, and the structure of the universe seem to be fine-tuned to the creation of intelligent life. Even an infinitely small deviation in any one of these parameters could have resulted in no universe and therefore no life.

4. Creation of physical laws cannot be thought of as a spontaneous process. There must be a lawgiver.

5. The second law of thermodynamics states that closed systems tend toward disorder, i.e., the entropy increases. But the universe and life on Earth show order rather than disorder. That leads one to believe that the whole system was predesigned.

6. Exclusion of supernatural forces or powers by science amounts to an ideological bias that hinders search for truth.

From the above arguments and the supporting literature, which is readily available, it becomes clear that the proponents of ID need, at the very least, recognition that their concept is a scientific theory. That, however, is not forthcoming from the scientific community. One of the problems is that no article in support of ID has ever been published in a peer-reviewed scientific journal. For scientists to provide any credence to ID as a theory, it must go through a rigorous peer-review process. So the ball is in the ID supporters' court. If the ID concept is to be considered as a scientific theory, it must first go through a scientific scrutiny by a jury of its peers (scientists with relevant expertise). If accepted and published by a science journal, it can rightfully argue to be a scientific theory. Until then it should be regarded as another version of the creationism doctrine.

In contrast to the lack of ID-supporting articles, there are numerous papers in peer-reviewed scientific literature that specifically refute the claims of ID. The reader may visit http://www.ncbi.nlm.nih.gov to access some of these articles. Therefore, instead of fighting ID vs. evolution battles in school board meetings or legal courts, the ID proponents should follow the scientific route and submit their findings for peer-review by scientists. That is the only way for any concept to gain recognition as a scientific theory.

While there is no scarcity of articles, pro or con, in support of ID in the general literature, neither side has a decisive proof as to whether there is or is not a creator of life. Darwin's theory does have the evidence and explanation for evolution of various life-forms, but it cannot exclude the possibility of a creator or an intelligent designer. The ID concept, on the other hand, looks at the complexity of life-forms and deduces the existence of an intelligent designer because it cannot explain it any other way. According to ID, life is so complex that it could not have emerged or evolved on its own without an intelligent creator or designer.

Is ID a Scientific Theory?

Although no one can prove or disprove the existence of an intelligent designer, it is quite straightforward to determine whether ID is a scientific theory or not. All one has to do is to apply established scientific criteria and see if ID meets them. While the reader is encouraged to apply these criteria or research the scientific literature on this topic, I have found no convincing evidence in support of ID. The reasons for my skepticism are stated below:

The claim by ID that IC (irreducibly complex) systems cannot evolve is without merit. All IC systems cited by the proponents of ID (e.g., clotting cascade, cilia, the immune system, flagella) have been shown in the literature to have evolved rather than been created as IC units. For an in-depth discussion of this point, the reader is referred to http://www.talkdesign.org/faqs/icdmyst/ICDmyst.html and other relevant articles published in science journals or scientific Web sites, such as http://www.ncbi.nlm.nih.gov/entrez/query.fcgi?.

Neither the concept of CSI nor the anthropic principle of *finely tuned universe* requires the necessity of an intelligent designer. Both have been well explained by the scientific theories of cosmic and biological evolution. For example, the creation and evolution of life were not an entirely random combination of atoms. Every random change or mutation is subjected to the laws of physics and the course of evolution dictated by the natural forces (i.e., strong nuclear, electromagnetic, weak nuclear, and gravity). Evolution is a much more dynamic and active process than passively stringing atoms and molecules together randomly to form the DNA molecule, for instance. Whenever a random change or mutation occurs, it creates a multitude of physical effects, one or some of which could result in a viable new form or expression of life. The statistics presented in the CSI argument as well as the anthropic principle are flawed because they do not take into account the influence of natural and evolutionary forces on random events.

There are some believers of ID who do concede that the universe and the life in it might have evolved as the scientific theories have shown. They also believe that there must be some intelligent being—a God—who created the laws that have governed cosmic and biological evolution.

There is considerable discussion in the scientific literature about how the laws of physics, chemistry, biology, and so on, which underlie all natural phenomena, came out to be what they are. A related question that has also been often asked is: why do the three fundamental constants—c (the velocity of light), h (the Planck's constant), and G (the gravitational constant)—have the values they do? If they were even minutely different, the universe and the life-as-we-know-it would have been impossible. Questions like these have given rise to ideas like the anthropic principle (chapter 8). Again, the interested reader is referred to the scientific literature to see if there is any merit to these arguments. However, one should not fall into the trap of believing that the lack of a convincing answer is evidence of an intelligent designer. Such a rationale is either a cop-out or a mindless leap of faith. After all, our forefathers deified all sorts of things they did not understand, like the sun, the wind, the lightning, and the fire, to name a few. So my suggestion to a reader who still might be confused is to open his or her mind with knowledge rather than close it shut prematurely with ignorance.

Another hotly debated issue is the *origin of the natural laws*. Some people, who believe in evolution as well as ID, argue that God created the laws, and the world has evolved according to those laws. Science maintains that the laws of physics came into existence as a result of the universe being born, that is, through quantum fluctuation of vacuum. The laws and their associated constants are properties of space and time. When the universe sprang into existence at the moment of the big bang, the space, time, energy, and forces were all one entity—the baby universe itself. As the universe expanded, the space-time resolved into space and time. Mass was created out of condensation of energy. The presence of mass distorted the space-time matrix to give rise to gravity. With the expansion of the universe from its point size, its binding forces got resolved into four forces that we know today, namely, strong nuclear, electromagnetic, weak nuclear, and gravity.

Natural laws seem to be derived from simplicity and homogeneity of nature. For example, the laws of conservation of energy, momentum, and electric charge are all expressions of homogeneity and symmetry of space-time. In fact, understanding the natural laws is a key to our understanding the universe.

The *entropy* argument of ID maintains that, while the second law of thermodynamics requires the entropy, or disorder, of the universe to increase with time, it is contrary to what we observe. For example, the present universe and the emergence of life on Earth show order rather than disorder. That means that an intelligent designer must have planned the universe that way.

The above entropy argument is without scientific merit. This argument would hold good only for a closed system of constant volume. Our universe is an open system as it has been expanding ever since it came out of its big bang singularity. In an open system, the entropy need not increase. Also, the entropy does not increase if the system as a whole maintains a uniform temperature, with no heat added or subtracted from it. In the case of the universe, most of the entropy is in its cosmic background radiation. With the expansion of the universe, the temperature of radiation has been dropping, thus maintaining constant entropy.

Although the entropy issue is not entirely settled, there is no evidence of any interference by an agent that is not subject to the natural laws. As is true of the whole concept of ID, its evidence rests on the thesis that if science cannot explain it, it must be from some intelligent designer—presumably God. Well, there are a lot of things that science cannot fully explain, but the scientists do not shut down the inquiry just because it cannot explain a phenomenon at a given time. The history of science is the history of man's ever increasing knowledge and understanding of the universe. Scientists do not look up to God for answers when they are confounded with the mysteries of nature.

I conclude the discussion of ID by saying that its proponents have produced no evidence that it is a scientific theory.

Is Evolution a Credible Theory?

The details of the theory of biological evolution were presented in chapter 8. Here we revisit the issue of whether Darwin's theory of evolution and natural selection is based on scientific evidence or speculation. As we have stressed earlier, the word "theory" in science does not mean an unproven idea as it is commonly thought of in layman's terms. I have often heard people say: "Evolution is just a theory, not a fact." Well, that is a misuse of the term *theory* when we apply it to evolution or any other concept that meets the scientific criteria of a theory.

As described in chapter 3, a scientific theory involves four steps in its formulation: (1) careful observation of the phenomenon; (2) tentative description or proposition, called hypothesis; (3) experimental verification and modifications, if necessary, of the hypothesis; and (4) formulation of theory from a consistently proven hypothesis. When a hypothesis becomes a theory, it is implied that considerable scientific evidence exists in support of the formulated principles. The modern theory of evolution is supported by evidence that meets the above criteria.

Skeptics of the evolutionary theory point toward what they call the "missing links." In other words, not all the evidence has been found that links the species together without discontinuities. Until the missing links are found, they argue, the evolutionary theory is just a theory, in the layman's meaning of the word, which is conjecture. Such an offhand dismissal seems to be a common stance of those who are either ignorant of the extensive evidence that exists or simply biased against any idea that runs counter to their religious beliefs. Nonetheless, for those who are inclined to an unbiased search for truth, there is no scarcity of convincing evidence. Some of this evidence was presented in chapter 8. For more in-depth study, the reader is referred to the relevant literature on evolution, which is in abundance.

In spite of the so-called missing links, the evolution theory has stood the test of scientific scrutiny for over a century and a half. Scientists may argue about details, but the basic idea that all living organisms, including humans, are the product of evolution that started billions of years ago on Earth is well accepted by the scientific community. Let us refresh the reader's mind with the evidence that overwhelmingly supports this evolution:

1. Fossils: The fossil records provide snapshots of evolutionary changes. The earliest traces of microfossils date back to 3.5 billion years. Because earlier forms of life were not able to fossilize, one cannot pinpoint the exact time when life began. However, from the available fossil data of the early Precambrian era, it has been inferred that life began over three billion years ago with the single-celled organisms, the prokaryotes, in the sea.

   One of the most powerful tools of paleontology is dating of rocks and fossils. Hundreds of thousands of fossils have been found, representing successions of life forms through time. These forms include prokaryote cells, eukaryote cells, multicellular plants and animals, invertebrates, vertebrates, amphibians, reptiles, mammals, primates, apes, early ancestors of humans, and modern humans. Sufficient numbers of intermediate forms have been discovered to allow paleontologists to construct evolutionary trees with branches and leaves. These carefully constructed graphic representations of evolutionary data point to our common ancestry—the prokaryotes.

   Currently there is an enormous amount of paleontological data available in support of the evolution theory. Critics may point out some gaps here and here, but the data continues to grow with new discoveries of fossils, and the missing links are being found even to this day. For example, the recent discovery of the *Tiktaalik roseae* fish fossil on Ellesmere Island in the Canadian Arctic in 2006[1] provided an important missing link from a period when sea-dwelling creatures moved on to dry land. This well-preserved fossil consists of a fish with lobe-like fins containing mammal-like bones and muscles. This fish, with body parts of a mammal, had been dead about 375 million years.

   Another recently discovered missing link is the prehuman species fossil, *Australopithecus anamensis*, which was unearthed in northeastern Ethiopia and posted by CNN on April 12, 2006.[2] This 4.2 million-year-old fossil fits in the human chain of evolution and allows scientists to fill in the gaps of how human ancestors evolved from one species to another.

2. Homologies: Related organisms, extinct or extant, show similarities in characteristics such as anatomy, cellular makeup, bones, embryological development, and vestigial structures. These similarities, called homologies, are consistent with the predictions of evolutionary theory. They suggest common ancestry.

3. Molecular evidence: The degree of difference between proteins and DNA or RNA in different species is closely related to the time of their splitting apart. For example, these molecules in humans are more identical to those in the chimpanzees than in the other vertebrates. Human genes are more than 99 percent identical with those of the chimpanzees, slightly less so for gorillas, and much less (about 80 percent) for the baboons. The evolutionary trees constructed on the basis of fossil studies are consistent with the molecular differences between different species. Interspecies relationships are well established.

4. Mutations: Mutations are one of the agents of evolutionary changes. Genetic mutation and genetic recombination can create new traits. If the new traits are advantageous to the organism in coping with the environment, a general mutation can give rise to an evolutionary change. Genetic science is the backbone of modern evolutionary theory.

5. Natural selection: The genetic variant of a species depends on the natural selection process to give it a chance to become established as a new species. Genetic mutation is a random process, but selection is governed by environmental conditions. Because the success or failure of a genetic variant cannot be predicted, evolution cannot be directed.

There are many other pieces of evidence in support of evolution. The reader is encouraged to consult the relevant literature or access specific information on the Internet.

Is there any doubt that evolution is a fact? None, so far as the scientific evidence is concerned. However, the evolution theory, like other theories in science, is not set in stone. It is being continuously studied and refined. As a theory it reigns as the factual story of life on Earth unless a different theory comes along and wins the scientific argument.

In spite of the fact that evolution is a widely accepted scientific theory of how life originated and diversified itself into countless forms, the question remains whether or not God had anything to do with it. By posing such a question, however, we find ourselves in a twilight zone where the natural merges with the supernatural. This may be allowed in a religion but not in science. Because, by definition, God is not subject to the laws of science, the question of his existence or his role in the creation cannot be investigated by scientific methods.

There are scientists, however, who believe in God as well as evolution. They take God's scriptures metaphorically rather than literally. For example, they do not consider the wording of Genesis 1 and 2 regarding creation as scientific explanations but metaphoric descriptions of God's powers as the creator. The mechanism of his creation is not specified. It is up to us to find out how he created the universe and all things in it, living or nonliving. Evolution theory, which is a human concept of how life evolved, does not concern itself with the question if God had anything to do with it. Therefore, there is no contradiction between evolution and the metaphorical interpretation of scriptural accounts of creation.

There are other scientists, probably the majority, who question the validity of anything that is based on the supernatural. They reject the very existence of a creator or designer. Evolution does not require any supernatural agent for it to operate. It's a fact of life based on scientific evidence rather than unsubstantiated stories told thousands of years ago.

Then there are the faithful who believe in the scriptures literally. To them the words in scripture mean exactly the same as words do in ordinary language. They do not want the divine message to be altered by far-fetched intellectual interpretations. According to those who believe literally in every word of the Bible, God created the universe in exactly the way he said he did—in six days, not in six seconds or six billion years. All species of animals and plants were created individually, not through evolution. Humans did not descend from monkeys. Humans were created by God in his own image, without undergoing evolution from any nonhuman species. To the literal believer, the evolution theory is incompatible with the scriptures and, therefore, should be regarded as false.

There are many variations of believers and nonbelievers of creation and evolution of life. The debate between science and religion will continue unless God, if he exists, reveals himself more than he did to Moses: "I AM THAT I AM."

CHAPTER 15
AT THE END OF QUEST

I was born and raised in the religious tradition of my family, and, like most people, I feel content with the religion I inherited. The religion of my forefathers is part of my humanity and a source of pride and comfort to me. I love my God more than anything in the world, and I often feel his love and caring in return. I thank him when I am happy, and I open my heart to him when I am grieving, for solace. I often beseech him to intervene when I see the world in trouble. When in doubt, I seek his guidance. When I show kindness to my fellow human beings, I get the feeling that I am doing his bidding, in addition to receiving my own personal satisfaction. To me, life without God is unimaginable and would seem boring to say the least. Without him I would feel alone and afraid, like a child in the dark.

But behind all these feelings of love for God, I also have a nagging doubt about his existence. It is possible that he is just a symbol, an imagined object of adoration and protection against my insecurities. Or, maybe he is just as real as the scriptures say he is, and, therefore, I should believe in him without proof. Only God knows what the truth is. Only God knows whether he exists!

I cannot avoid the feelings of doubt about God, nor do I make any attempt to resist those feelings. I have come to believe that my God, if he exists, would not be displeased with me having these doubts. After all, he has not revealed his existence to me. He would not want me to shut my mind to rationality or have a blind faith in him with no questions asked. I love God for my own emotional needs, not his, and I use my own mind, not his, to figure out what is right and what is wrong. I don't think my God is slighted when I doubt his existence. I don't think he is slighted when I use my brain as best as I can to figure out things instead of some irrational dogma or a doctrine promulgated in his name. I feel that he would be pleased if I love his creation rather than worshipping him, glorifying him, or singing his praises. My God would want me to be a good human being, using my own abilities to my full potential rather than spending all my life figuring out what he would want me to do or not do.

When I ponder the innumerable attributes of God, I get the feeling that they are the extension of human goodness. A psychologist might say (not to be taken literally) that God is the creation of man in his own image of what is perfect. God is what man would like to be. God represents the soul of man. In my personal morality and ethics, God is my conscience.

In spite of my boundless love and admiration for him, I often felt in the past that I did not know God beyond what I learned about him from my family, friends, teachers, and the scriptures. In a way, I had to come to love an imaginary being called God. My quest for "the real" God started when I was introduced to astronomy and cosmology. My search for him has intensified since I retired from my professional job in 2001 (probably because I had more time to think and reflect about him).

The initial objective of my quest was to learn about God's role in the creation of our universe. With my background in physics, I first turned to cosmology for answers about how the universe came into being and how it grew up to be what it is today. Until then, I had very superficial understanding of the big bang model, gravity, general relativity, and other astronomical facts associated with the universe. Cosmology is a vast subject, and it is quite daunting to undertake an in-depth study of it. I narrowed my focus to the scientific logic behind what caused the universe to come into existence. I must admit that throughout this study, there was this nagging question of a creator in the back of my mind. I was looking for the slightest clue that would suggest a cause other than the spontaneous creation. I used to think, like most people would, that nothing happens without a cause. But as a physicist with some familiarity with quantum mechanics, I was able to understand that the universe could very well have come into existence by itself. Although it is not 100 percent conclusive, and nothing is in the field of science, cosmological theories and astronomical data strongly suggest that that is what happened in the case of our universe.

Cosmology is a fascinating subject, and I thoroughly enjoyed studying it. However, I could not find evidence in cosmology or astronomy that would be suggestive of some agent or force outside the universe that might be responsible for its creation. This statement should not imply that God did not create the universe. What it says is that science does not have any evidence that there was a causal agent for the creation of the universe. In other words, science has no way of proving or disproving a supernatural phenomenon. God is outside the realm of science.

I also gave considerable thought and study to the anthropic principle that has been proposed to suggest that the fundamental constants (c, h, and G) are so fine-tuned to the creation and operation of our universe that it must be the work of a supreme being. I did not buy that. I did not find any convincing evidence to justify that claim. You cannot prove God by default!

One of the most difficult things for me to understand was how the universe got started by itself from nothing. It took me a long time to make sense of the physics of "nothingness." Next, I tried to make sense of the concepts involving quantum fluctuations of vacuum and the creation of positive energy from zero energy. Einstein's general theory of relativity has always been a challenge for most physicists to understand. You can easily give up under the heavy weight of his mathematics. His explanations for laypeople are not that easy to understand either. However, there are plenty of books written by others that are very helpful in understanding his theory.

Of all the forces of nature, gravity is the most difficult to understand. Yet, it holds the key to our understanding of the universe. Thanks to Newton and Einstein, we had a very good start. Quantum gravity holds great promise in solving some mysteries of the universe, especially the mystery of creation and its subsequent expansion. I am waiting for the day (I hope I will still be around) when a unified theory is developed that will explain the whole thing, especially how it all began. The string theory is still in its infancy, but who knows?—it may turn out to be the one we have all been waiting for—*the theory of everything.*

The study of cosmology makes you realize that there is nothing outside the universe—no space, no time, no energy, no matter. If God exists outside this universe, then he must not be of the same substance this universe is made of. He is neither energy, nor force. He does not occupy space, nor is any time associated with him. In that sense, he meets the definition of what the scriptures call God. However, that does not help us understand him, scientifically I mean. So what is man to do? Should we just believe in what Moses, Jesus, and Mohammad have told us about him—which is really not much to go on in scientific terms? If you are looking for any physical evidence, you will soon come to a realization, as I did, that there is no scientific evidence, either in science or in the scriptures that God actually exists.

How about nonscientific evidence? That gets to be tricky because we have to define what nonscientific evidence is. Nonscientific evidence simply implies that the evidence exists outside science, such as in the scriptures, which cannot be proven using the scientific method. That means the evidence for God's existence is not going to be found in science, although the scientific claim of spontaneous birth of the universe does chip away at the religious concept of a creator. What about miracles?

Numerous miracles are recorded in the scriptures and in folklore that are impossible to verify for accuracy of reporting, let alone for any scientific evidence. Therefore, it would be fair to say that nonscientific evidence is no evidence at all. Opinion, intuition, gut feeling, belief, or philosophical speculation does not constitute evidence—nor does a revelation from God without scientific proof. The question whether God exists or does not exist must involve science, or it should not be argued. Faith is not debatable.

After convincing myself of the fact that science has no evidence for God's existence or nonexistence, I turned to religions in my quest for God. My thinking at this point was that God is a religious concept, and, therefore, one should study religion in order to understand God. There are thousands of religions, sects, and denominations, each with a slightly different take on God. Which one should one turn to for answers? It made it all the more difficult for me because I was determined to do an unbiased search for truth. So, I decided to look at all the major religions and find out what their understanding was specifically regarding the nature of God.

I soon found out that, just like science, religion is not easy to study or master in a few years' study. However, I was greatly helped by my general familiarity with several religions that I had acquired on my own as a matter of curiosity and interest. So I embarked on my quest with great confidence and anticipation. I often talked to my family and friends about this project. My frequent motto was: "I want to get to the bottom of it."

By focusing narrowly on the question of God, I was able to hasten my search a little. I quickly realized that Eastern religions (e.g., Hinduism, Buddhism, and Confucianism) are more akin to the concept of Ultimate Reality than that of a personal God, as espoused by Western or Abrahamic faiths. This difference in concepts, although very interesting, is not an argument for who is right or wrong. But it is definitely an indication of the fact that the concept of a supreme being, such as God, is not unique to one people, one culture, or one civilization. It is very much related to time, history, and geography.

One could look at the conceptual differences about God in two ways: (1) there is one Supreme Being, but he revealed himself to different people at different times, and thus his message or revelations are different or have changed accordingly; and (2) God is just a human concept, in which case the understanding of God is bound to change with the advancement of civilization. God is not a static concept, whether revealed or imagined.

Anthropologists and sociologists have done extensive studies on the origin and development of different religions. There is strong evidence that the concept of

God has evolved with the evolution of human thought. From my studies I have learned that no religion can claim exclusive privilege of God's knowledge or revelation. The conceptual understanding of God varies not only between different religions but also between followers of the same religion. Therefore, which God you should believe in is a matter of personal choice and conviction.

I say religion is personal because it is a system of beliefs, not a scientific theory. You cannot impose your beliefs on others because their beliefs may be different than yours. You cannot claim that you are right and others are wrong because you have no proof for either one. The only acceptable proof in this case would have to be a scientific proof because that is the only common currency of human rationale. But such a proof is not forthcoming. Let us face it. Beliefs are just beliefs; they are not proofs.

That brings me to the question of *exclusivity* which is prevalent among most religions. Each religion preaches its own truth, and all truths are exclusive by definition. However, not all followers of a religion are exclusivists. Each religion, like other human institutions such as politics, economics, and law, is divided into conservatives and liberals, with many shades of gray in between. Religious conservatives tend to be exclusivists—they believe that their religion is the only way to salvation. Religious liberals, on the other hand, believe in the truth of their own religion but do not rule out salvation for followers of other religions or even atheists, provided they lead their lives in accordance with the general principles of human ethics and morality. As a result of these differences, most religions are split into sects, denominations, and cults, with their followers further split between conservatives, liberals, and those in between. It is amazing to realize that religions reflect politics quite closely and vice versa.

Religious exclusivity is understandable in the sense that new religions are created on the basis of being the only truth. If others were equally true, then what is the point of claiming separate identity? However, the extreme view of exclusivity, in which followers of other religions are automatically condemned to hell, poses problems with the concept of God as the creator and the most compassionate. How could this God condemn entire cultures that practice different religions or have no religion at all? For example, if we take the following biblical verse literally, the majority of the people in the world today would not attain salvation, that is, they would not enter heaven or would be condemned to hell or meet similar punishment:

> Jesus saith unto him, I am the way, the truth, and the life. No man cometh unto the Father, but by me. (John 14:6)

There are many other similar exclusivity statements in the Bible as well as in the scriptures of other religions. If these were taken literally, no one would likely escape God's wrath, except the lucky ones who happen to follow the right religion. Since most people are born into a religion rather than adopting one later by choice, it seems our fate is practically sealed at birth.

Literal beliefs of exclusivity can give rise to extreme forms of exclusivism. Radical exclusivists not only believe that their religion is the only true religion and that all others are false, but they make frequent public pronouncements to that effect to make their point. In a pluralistic society with diverse religions, this could breed intolerance to other faiths. History is replete with sectarian violence and religious persecution caused by radical exclusivism. More extreme forms of exclusivism have given rise to brutal wars, crusades, crimes against humanity, and genocide.

Exclusivists who regard their scriptures as infallible and base their beliefs on a literal interpretation of the texts also tend to be *fundamentalists*. Fundamentalism, which exists in almost all religions, is a movement to return the believers to the defining or founding principles of the religion. It is antimodernist as it perceives modern society in a state of religious and moral decay.

Religious fundamentalism as a personal belief seems to be benign on the surface, but when it is preached and promoted openly in the general public, it can breed hatred and intolerance toward nonconformists. The fundamentalists are not only concerned with their own personal beliefs; they also demand conformity to their beliefs from the rest of society. The following examples illustrate the aggressive nature of radical fundamentalism:

> I want you to just let a wave of intolerance wash over you. I want you to let a wave of hatred wash over you. Yes, hate is good ... Our goal is a Christian nation. We have a Biblical duty; we are called by God, to conquer this country. We don't want equal time. We don't want pluralism.[1]
>
> —Randall Terry, founder of Operation Rescue
> Reported by the *News Sentinel*, Fort Wayne, Indiana
> August16, 1993

> We are importing Hinduism into America. The whole thought of your Karma, of meditation, of the fact that there's no end of life and there's endless wheel of life, this is all Hinduism ... The origin of it is all demonic. We

can't let that stuff come into America. We have got the best defense, if you will ... a good offense.[2]

—Pat Robertson, *700 Club* television program
March 23, 1995

When fundamentalism combines with politics, it can give rise to theocracy, dictatorship, or fascism. Because religion derives its powers from God, it is prone to being used as a tool to advance fundamentalist ideologies. The opposition is often demonized as being infidels, atheists, or godless liberals.

In my study of religions, I have found that fundamentalism, combined with politics, is a negative force that tends to divide communities and cause a multitude of problems ranging from sectarian violence to acts of terrorism. So a nation like ours, with plurality of cultures, religions, races, and ethnicities, must be on its guard against religious fanaticism. We need to let love, compassion, and justice have a higher priority than the fundamentalists' ideology of exclusivism and doctrinal conformity.

Most fundamentalists are against *separation of church and state*. The word *church* here means any religious establishment. In the United States, the principle of separation of church and state is derived from the establishment clause of the First Amendment to the U.S. Constitution, which states: "Congress shall make no law respecting an establishment of religion, or prohibiting the free exercise thereof." This great principle protects both the state and the religion from each other—from religion's interference with government and government's interference with religion. The above statements by Randall Terry and Pat Robertson reflect the fundamentalists' mindset. To them there is no other truth but their truth.

There may be no actual roasting or flogging of religious opponents in this day and age, but the other evils such as discrimination, abridgement of human rights, and criminalization of innocent citizens could be perpetrated on those who might have different religious beliefs. It should be understood that the separation of church and state is not a secular principle. Rather it is an affirmation of the principle that "Congress shall make no law respecting an establishment of religion, or prohibiting the free exercise thereof."

Enlightenment

When my serious quest for God began in 2001, my main question was, "Does God really exist, or is it just a human concept?" I also faced some difficult questions about the origin of the universe. Who or what created the universe? And, if it was created by God, then who created God? Other related questions were equally daunting. Is there life after death? Is there a judgment day on which God will judge everyone for his or her actions and beliefs? Is God the source of all religions, or does each religion have its own God or gods? Finally, to shorten the never-ending list, what is life, and what is its purpose?

I soon realized that all these questions were interrelated and that the question of God was at the top of the list. Furthermore, you cannot learn anything about God without knowing the answers to the other questions down the list. So, I decided to study the universe in order to learn about the creator. I researched major religions to get some insight into their concepts of God. My plan was simply to look for evidence, scientific or otherwise, that would convince me about the reality of God. Although I knew very well the difference between science and faith, it did not make sense to find answers about God in one with the exclusion of the other.

Before I started the quest in earnest, I made the following promises to myself:

1. I shall cast away all my prejudices that I could identify. The quest must be an unbiased search for truth, as much as humanly possible.

2. In case of competing ideas, concepts, or theories, I must approach them with an open mind.

3. I must rely on my mind to understand the evidence in order to form a conviction.

4. I must commit myself to an honest pursuit of truth to the best of my abilities and conscience.

At the end of my quest, lo and behold, I FEEL ENLIGHTENED!

Although the word *enlightenment* means different things in different contexts, I feel closest to the definition in Wikipedia, which describes enlightenment as "being illuminated by acquiring new wisdom or understanding."[3]

I would like to share my "new wisdom and understanding" with my readers who probably have questions about God similar to the ones I had at the beginning of the quest. After reading this book, hopefully the reader will understand the logic of my conclusions and the reasons for my enlightenment.

Although my personal quest for truth about God and nature continues, I have attained my own version of "nirvana" regarding God, humanity, life, ethics, morality, and the origin of our universe. In this state of "blissful rapture," I enunciate the following insights that I have gained as a result of this quest:

God: There is no scientific evidence, direct or indirect, that there is a God. Nor is there a way to prove or disprove the existence of God through science. This is true of all supernatural phenomena because they are not amenable to the scientific method of inquiry.

Mankind has learned about God through prophets, who were human beings, and through scriptures, which were written by human beings. Religions claim that prophets were messengers of God and that the scriptures were either revealed by God or inspired by God. However, just like God, there is no scientific evidence to prove or disprove that prophets communicated with God or that the scriptures are God's words transmitted through revelation or divine inspiration.

Although science is helpless in providing evidence of God's existence, it has the tools to investigate the origin of the universe, including life. Modern cosmological theories such as the big bang model (chapter 6) have already provided a few answers with regard to how the universe was created and how it evolved. Some of these findings contradict the scriptural accounts, if the latter are taken literally. Is that a proof that the scriptural stories are not true? I would say yes only if we go by the literal interpretation. But who knows if these accounts are deliberately metaphorical or if it is God's way of teasing our mind to solve the mysteries of the universe ourselves. Then again, it would be pure conjecture on my part to think that the scriptures are all metaphors.

Setting aside for the moment the question of whether the scriptures are divine in nature or not, most people, including myself, consider all scriptures to be great literary pieces. They are the treasures of human heritage and should be treated with reverence. To analyze the scriptures for scientific accuracy is fruitless because, like great art or literature, they deal with human emotions rather than scientific facts. Any contradiction between science and scriptures should be attributed to this chasm that exists between science and art or literature. Word-for-word translation and interpretation of scriptures run the risk of their being refuted by science and losing their literary value. Literary truths are not the same as metaphorical or scientific truths.

Just because we do not have scientific evidence of God's existence or his revelations to mankind, should we then not believe in God? My answer is: it is all up to you. If you decide not to believe in God, then there should be no hostility shown to you by those who believe in him. Not believing in God in all earnestness may be an error, but it is neither a sin nor a crime. The Qur'an makes it clear:

> Let there be no compulsion in religion: Truth stands out clear from Error: whoever rejects evil and believes in Allah has grasped the most trustworthy hand-hold, that never breaks. And Allah hears and knows all things. (Qur'an 2:256)

If you believe in God, then it should also be clear in your mind that your belief is based on faith, not scientific evidence. And precisely for that reason you have no right to impose your belief on others. You can preach, persuade, or induce others through reasoning to believe in God, but "let there be no compulsion."

Another important point for the believer to understand is that his or her belief is a matter of personal faith, not a scientific rationale. That means that a nonbeliever's disbelief is also a matter of his or her faith. So both the believer and the nonbeliever have equal standing in their earnestness of their convictions. Only God, if he exists, can be the arbitrator of the real truth, and no human knows the mind of God. Therefore, it is anti-God in principle to punish, harass, or discriminate against those who do not believe in him. No one should play God's vigilante lest they incur his displeasure.

A belief has a built-in uncertainty. Unless it is a scientifically verifiable fact, the belief is inherently uncertain just by virtue of it being a belief.

Faith in God transcends scientific reality. It may represent a higher truth beyond the reach of science, but it is a faith nonetheless. Because it is a belief with no scientific evidence, your chance of being right is the same as your chance of being wrong. Therefore, there is no rational basis for being 100 percent right about something that is in fact an intuitive guess. To lose sight of this reality is to be blinded by irrationality.

The ugly fact of uncertainty, however, should not spoil your personal relationship with God. If you believe in God, then let it be an unwavering faith. It doesn't make sense to believe in God by following Pascal's Wager:

> It is a safe bet to believe in God whether He exists or not. Because if you believe and He does not exist, there is no harm done. But if you do not believe and He does exist, you are surely going to hell.[4]

As a believer, your love for God need not be marred by doubts. The only time you need to acknowledge the reality of uncertainty is when you are dealing with other people and their beliefs. As a man of faith, you must respect other faiths because, who knows, they may be right and you may be wrong. Since no one can prove the correctness of one's faith, the rational thing to do is to accept the reality of this uncertainty.

If you happen to be an atheist, however, your uncertainty of belief in God has changed into a certainty of denial. An atheist does not believe in God because he or she does not believe in anything without evidence. In that case no scientific evidence exists or will ever exist of God's existence. So the atheist's denial is consistent with his or her belief system (i.e., no belief without evidence).

Religion: Belief in a particular God or in an Ultimate Realty does not constitute a religion by itself, although it may be a focus or nucleus of religious activity. Religion is an expression of faith through worship, ritual, and conduct, often involving a code of morality and a philosophy of life.

Because most people inherit their religions through their upbringings, it is important to recognize that the religion of birth may or may not be the one true religion. Accepting your religion of birth as the true religion is your choice. If it meets your emotional and spiritual needs, then it is the right religion for you. But do not lose sight of the undeniable fact that there are billions of people who are just as convinced about the truth of their religions of birth as you are of yours. So the truth of a religion is a personal belief rather than a universally acknowledged truth. Religious truth is in the mind and heart of the believer.

Worship and prayer: An important part of a religion has to do with the rituals of worship and prayer. You worship God to glorify him, praise him, exalt him, and honor him for all his attributes. Scriptures are full of praises for God that God himself instructs the believers to utter frequently to beseech his favors. You worship God to please him so that he will reward you and protect you from harm. You ask God's favor by praying—letting him know your feelings and wishes. You invoke God for his blessings, inspiration, and support in your daily affairs as well as in emergencies. When you are sick, you ask him to make you well. When something good happens to you, you credit him and thank him. When a loved one dies, you do not blame him but pray to him to save the deceased's soul. Out of the goodness of your heart, you pray for others just as you pray for yourself. Sometimes you curse your enemies by wishing God to punish them or destroy them. Although you never know if your prayers would be answered, you feel good about asking his favor. This is a typical relationship between a believer and his or her personal God.

Some people consider worship and prayer pointless because the God that they believe in is omniscient and knows everything in their hearts. He could grant your wishes without you asking. Also, he does not need any praise or glorification unless he is vain and self-conceited. So why do believers worship and pray? Skeptics think that these are essentially manifestations of man's selfish nature augmented by his religious neurosis and delusions about a father figure who is believed to be omniscient and omnipotent.

Although one would think that God does not need any worship or prayers to massage his ego, the believers do them because the prophets and scriptures have said that this is the way to access God's compassion. In other words, a personal relationship with God needs to be established through worship and prayer in order to receive his blessings. In the believer's mind, it is a two-way street. This is why he is called a personal God.

Because God knows everything in your heart and mind, you do not need to worship or pray in public unless it is a community ritual where the faithful gain strength from each other. Since faith is a personal matter, you do not need to make a public display of it. Jesus Christ makes this clear in his Sermon on the Mount:

> And when thou prayest, thou shalt not be as the hypocrites are: for they love to pray standing in the synagogues and in the corners of the streets, that they may be seen of men. Verily I say unto you, they have their reward.
>
> But thou, when thou prayest, enter into thy closet, and when thou hast shut the door, pray to thy Father which is in secret; and thy Father, which seeth in secret shall reward thee openly. (Matt. 6:5–6)

Heaven and hell: All major religions predict certain rewards for its adherents and punishments for those who are defiant of God or the Ultimate Reality. Descriptions of heaven and hell are essentially the same in all religions, although variations exist in the details or with regard to the recipients of the reward and punishment.

You, the reader, may be content with whatever God has prescribed for the good and the evil. But I have problems with the very concept of reward and punishment. I am perturbed by the scriptural descriptions of hell. I do not believe that the God I love—the God of my conscience—would be so cruel as to burn the sinners in an eternal fire. I can accept his not raising the souls of the sinners after death and depriving them of the heavenly rewards that await the believers. But to torture them for eternity is cruel and unusual punishment even from human standards, and we humans do not claim to be anywhere as forgiving, just, or compassionate as God is.

Hell is indeed a torture chamber. The following descriptions are presented as samples of sinners' punishments (reader's discretion is advised):

> He went from there to the east. There men were dismembering one another, cutting off each of their limbs, saying, 'This to you, this to me!' He said: 'O horrible! Men are here dismembering one another, cutting each of their limbs!' They replied, 'In this way they have treated us in the other world, and in the same way we now treat them in return.' He asked, 'Is there no expiation for this?' 'Yes, there is.' 'What is it?' 'Your father knows it.' (Satapatha Brahmana 11.6.3)[5]

> To begin with, the wardens of hell subject the sinner to the firefold trussing. They drive red hot iron stakes first through one hand, then through his chest. After that they carry him along to be trimmed with hatchets. Then, head downwards, they trim him with razors. Then they harness him to a chariot, and make him pull it to and fro across a fiery expanse blazing with fire—the flames leap and surge across, and fill it through out. (Theravada Buddhism. Majjhima Nikaya, in Conze, Buddhist Scriptures, 224–5)[6]

> And they shall go forth, and look upon the carcasses of men that have transgressed against me: for their worm shall not die, neither shall their fire be quenched; and they shall be an abhorring unto all flesh. (Isa. 66:24)

> But the fearful, the unbelieving, and the abominable, and murderers, and whoremongers, and sorcerers, and idolaters, and all liars, shall have their part in the lake which burneth with fire and brimstone: which is the second death. (Rev. 21:8)

> Nay, they deny the Hour (of the judgment to come): but we have prepared a Blazing Fire for such as deny the Hour:

> When it sees them from a place far off, they will hear its fury and its raging sigh. And when they are cast, bound together, into a constricted place within, they will plead for destruction there and then! (Qur'an 25:11–13)

There are many other descriptions of hell in various scriptures (about 162 in the Christian scriptures alone). Most are characterized by fire while some describe other forms of torture. I don't know the sensibilities of ancient people to whom these scriptures were revealed, but these descriptions of hell would be revolting to most people today. I have a feeling that if the scriptures were read carefully and taken literally, these torture chambers of God would shake the foundations of faith in most believers. The concept of hell may have been a deterrent against sin for the primitive man; it certainly isn't effective for the modern man.

Morality and ethics: Morality refers to principles of right and wrong in human conduct. Ethics is the system or code of morals. The two terms are often used interchangeably. Ethics may also be called moral philosophy.

Human conduct or behavior may be characterized as moral or immoral depending on whether or not it conforms to the generally accepted standards of goodness or rightness in conduct or behavior. Moral standards may be based on criteria set collectively by a society, established religious doctrines, philosophical reasoning, or scientific rationale. As these criteria change with time, so do the moral standards. It is important for us to understand that as human civilization evolves, morality also evolves to stay relevant to our contemporary needs and moral values. Even though certain moral standards may be relatively constant with time and universally agreed on, their correct application to specific situations at specific times may cause controversy and debate. Resolution of such debates may bring about a revision or reaffirmation of the existing standards of morality.

Although there are some moral standards that are almost universal (e.g., the Ten Commandments or similar versions in other religions), not all cultures in human history had the same moral systems or philosophy. Human sacrifice, infanticide, cannibalism, slavery, and so on were not considered immoral in some cultures of the not-too-distant past. Religion had a lot to do with making reforms in the existing morals and introducing some new ones. Religion was effective in bringing about changes in morals because it derived its power and absoluteness from God. But not all religions represented the same God and therefore not the same morality. Even in the Abrahamic religions, there are differences, although not very basic, in some moral guidelines such as prohibitions of eating pig's meat, drinking alcohol, polygamy, and divorce. In addition to these differences, there is no way of proving superiority of one moral code over another because each religion claims absoluteness of its own truth.

Should morality be considered relativistic? Are all moral judgments derived from religious sources, cultures, events, or people? Is there an *objective morality* that could be based on scientific logic?

Answers to the above questions are difficult to find because morality is a subject that is not easily amenable to objective analysis. In an article, Eugene Khutoryansky provides a good discussion of these issues and argues that morality can in fact be based on or justified by scientific reasoning. His position is that morality is linked to human emotions, such as suffering, fear, compassion, love, hate, envy, and jealousy, and by scientifically analyzing human emotions, the basis for objective morality can be found. For example, according to Khutoryansky:

> It is only by experiencing suffering ourselves that we can come to understand what suffering is and the fact that it is wrong to inflict it needlessly. This, however, should not come as a surprise. All science is based on observation.[7]

Although *scriptural morality* derives its authority from God, it is problematic for the same reason. The author of the moral code, namely God, is not responsible to the people who have to live by it. You are commanded by God to follow his moral code or else. Although there is some wiggle room in terms of theological discourse or scholarly interpretation within a given religion, interreligious differences are usually not debatable. So the religion-based morality is just as correct as the God who revealed it. For that reason, a country with religious diversity would be well-advised to stay away from making laws based on religious morality if it professes impartiality toward all its citizens—people of different faiths, including agnostics and atheists. The objective moral principle in this case would be to not impose your morality on others just like you wouldn't like others to impose morality on you. Morality should be based on scientific reasoning rather than supernatural beliefs of one religion or another, which are neither objective nor universal.

The above discussion should not be taken as a criticism of religion as a source of morality. Historically, religions have been in the forefront of bringing morality into the world. Prophets and scriptures brought light into the dark ages of human civilization. Through morality, man was transformed from being a savage animal to a civilized human being. However, we have come a long way from the era of prophets and scriptures. This is the age of science. We should guide ourselves through scientific reasoning while respecting our roots and each other's faith, whatever that may be. This is the age of enlightenment.

CHAPTER 16
LIVING WITH FAITH AND REASON

Environment is a sculptor … a painter. If we had been born in Constantino-
ple, then most of us would have said: 'There is no God but Allah, and
Mohammad is his prophet.' If our parents had lived on the banks of the
Ganges, we would have been worshippers of Siva, longing for the heaven of
Nirvana. As a rule, children love their parents, believe what they teach, and
take pride in saying that the religion of mother is good enough for them.[1]

—Robert Green Ingersoll,
American politician and lecturer (1833–1899)

I believe many people have gone through a similar process and have kept their
religion of birth as their guiding light throughout their lives. Some might have
had a change of heart, converted to another religion, or given up on faith altogether. Is there any way to make everybody follow the same path? Is there any
religion that can prove superiority over all others? Is your religion of birth the
correct religion? In the absence of an objective test, like the scientific method,
there is no right answer to any of these questions. So my dear reader, accept your
fate and forget about the futile search for a really true religion. Be contented with
what you have got, and if you like it, keep it and love it. If not, change it or give
it up altogether. As a free man, you don't have to get shackled by any particular
religion, creed, or philosophy. The only tool you can rely on in making your
choices is your "God-given" mind. And do not think for a moment that "your
God" would like you to believe in anything mindlessly. Do not take a chance on
a blind faith. Do not take a shot in the dark. Use your head. If you are convinced
in your mind and heart that your religion is the correct one for you, then love it
and cherish it. Call it a gift from your God.

The same reasoning should apply to alternative philosophies like humanism,
secular humanism, atheism, and agnosticism. Some people just cannot live with
or cherish supernatural beliefs. Rationality of their beliefs is their prized asset.

Having a liberated mind is also a gift, irrespective of whether you believe in God or not.

Faith and reason can coexist, but you have to understand the limitations of each. Through faith, you can reach transcendental truths that are not accessible through scientific reasoning. By the same token, you learn about natural truths through science that are not revealed or explained sufficiently in the scriptures.

Mixing faith and science without distinction, however, is fruitless. It is not only futile; it could contaminate both, thereby hiding the truth that you are seeking. A natural truth revealed by faith must satisfy the criteria of scientific method before it is labeled as such. In the absence of scientific evidence, it is a belief, intuition, or just a leap of faith.

If a conflict occurs between science and religion over a natural phenomenon, the truth is on the side of evidence. Because empirical evidence is verifiable and faith-based evidence is not, the former legitimizes its claim to that particular truth. Those who deny scientific truths because of conflicts with their faith are outside the bounds of rationality. Although they have the right to believe that their faith transcends science, they do not have the right to impose their beliefs on others. Better yet, they should realize the limitations of their beliefs and not deny scientific evidence just because it conflicts with their faith. Scientific evidence can be refuted only through the scientific method.

Principles of Enlightenment

After a long and arduous search for truth, I feel enlightened about what I have learned about the universe, God, religion, and humanist philosophies. I feel that anyone can undertake such a journey and thus get rewarded. The only precondition is that the seeker of truth must cast aside prejudices as much as humanly possible. It is the prejudice that kills the chances of finding the truth. Once you break out of your inherited beliefs and prejudices, you are intellectually free. With a free mind, you can acquire new wisdoms and untainted understandings of things from the great reservoir of human knowledge that exists today.

From the same reservoir, I have extracted some fundamental truths that I regard as my guiding principles of life. I would like to share these with my readers with the hope that they, too, would be inspired to launch their own quests and utilize their own creative minds in the fulfillment of their goals in life.

Science:

1. Science is the most valuable human asset. It is through science that we are able to learn about the fundamental truths of nature and the physical world.

2. The scientific method is the only way to establish the truth of a natural phenomenon.

3. Some truths may lie outside the reach of science. Unless they are verified by science, they are just opinions or beliefs.

4. All thought, feelings, memory, and self-consciousness originate in our brain. The regions of the brain that are responsible for mental activity are collectively called mind. Your brain contains both the hardware and software of your mental activity.

5. Our mind is not infused with some soul or spirit linked to some supernatural power. Mind is nothing but a product of the brain's activity.

6. Self-awareness or consciousness is an affirmation process mediated by the brain at any time to establish recognition of itself and the surroundings. Thoughts, feelings, action plans, and self-consciousness are all states of mind brought about through physiological activity of the brain. The mental experience is the result of interactions among billions of neurons within the brain.

7. There is no scientific evidence for a separate entity, called "soul," that is associated with human existence. However, if there is a soul, we have no way of detecting it. Like God, the existence of the soul cannot be proved or disproved by science.

8. There is a credible scientific theory that our universe sprang into existence spontaneously by a quantum fluctuation of vacuum (a state of nothingness with no space, no time, no matter, no energy). Our earth is the product of cosmic evolution that began about 13.7 billions years ago.

9. All life on Earth originated from a single point. All species have descended from a common ancestor—the progenitor organism.

10.   There is strong scientific evidence that human beings evolved on Earth from the first cells of life, called prokaryotes, some three billion years ago. These single-celled organisms were the result of atomic and molecular interactions on the surface of the earth under the right chemical and environmental conditions. There is no doubt that the first life was created by the same laws of physics and chemistry as those that govern the nonliving things. There is no evidence of an "intelligent designer" of life.

11.   The basic principle of biological evolution is well stated by Philip Whitfield:
      The favoring of changes that are helpful in life's actual context seems to suggest that some guiding force is pushing evolution in a preordained direction. This is a recurring and seductive misconception. There is no external intelligence in life that guides evolutionary change. Evolution is not moving toward a predetermined goal nor is natural selection equipped with a compass. Since evolutionary change is born of random variation and local contingency, the only direction it can take is decided by those randomly generated changes that happen to be successful in specific circumstances.[2]

12.   There may or may not be absolute truths or an ultimate reality. The scientific truth is conditioned by the validity and durability of its scientific evidence. They are not everlasting absolute truths, credos, or dogma.

13.   Scientific theories, by definition, are neither conjecture nor speculation. They are backed by considerable evidence in support of their validity. Without the supporting data, they would not be called theories. The only way to refute a theory is to follow the scientific method of disproving it.

14.   The theories of the origin and evolution of the cosmos and life on Earth are scientific theories. You believe in them on the basis of scientific evidence until and unless they are contradicted by new scientific theories or facts.

God:

15.   Science is unable to prove or disprove the existence of God.

16.   I believe in God as a matter of personal faith. I feel that my belief in him fulfills my emotional needs. I also believe that God has guided humanity from time to time through inspired men, prophets, saints, and reformers. I

also realize that my belief in God is intuitive and unverifiable. I recognize that it is a willful exception to my rationality.

17. I consider it immoral (and against the wishes of the God that I believe in) to force others to believe in God or spread the faith by coercion.

Religion:

18. In world history, religions have been both constructive and destructive. On the positive side, they have stressed moral conduct in order to realize the highest values of life. On the negative side, they have caused wars, persecution, and misery for those who disagreed or did not believe.

19. Religious doctrines are introduced to solve the problems of human living. However, once established they rarely undergo reformation through an open debate. Because of that, they tend to lose their significance with time. Resistance to change is often opposed by those who consider religious doctrines as ordained by God and therefore everlasting and unchangeable.

20. Exclusivity and literal conformity to the holy scriptures have given rise to fundamentalist movements in almost all religions. Fundamentalism does not bode well for either the religion or the people it serves.

21. Any religion that cannot furnish adequate social goals or personal satisfaction to its believers should reform or face extinction.

22. Any religion that conflicts with the natural truths should reform or run the risk of losing its credibility.

23. Separation of church and state is essential for both to survive and prosper.

24. There is no scientific evidence that life could be reconstituted after death. The concepts of reincarnation, resurrection, immortal salvation, and eternal damnation are illusionary and have no basis in reality. The idea of reward and punishment after death might have worked for people in the ancient past, but it is meaningless or irrelevant to the needs of present-day civilization.

25. Religions are fountainheads of ethical teachings that should be preserved as human heritage. Those features that foster exclusivity, discrimination, hatred, sectarian conflict, or holy wars must be rejected.

26. Religious traditions that keep humans from realizing their full potentials and responsibilities should be discarded.

27. All humans have the fundamental right of choosing and following their own spiritual, religious, humanist, or secular paths in life.

28. We must respect the rights of people to hold diverse beliefs if we want others to respect our beliefs.

29. Human beings are responsible for their own actions—good or bad. Our destinies are in our own hands.

Morality:

30. Morality must be based on human experience and the understanding of human behavior. It must be relevant to the needs and desires of people, individually and in social settings.

31. Certain actions are inherently right or wrong. These are universal moral values that are prevalent in all cultures, religions, and societies.

32. It is possible to judge the morality of an action through rational and scientific reasoning. Such a morality is called objective morality because it has the same rationale regardless of religion or culture.

33. Belief in God or religion is not necessary in order to follow objective morality. No religion, culture, or group has a monopoly on virtue.

34. There are certain moral issues that are controversial. Rationality, scientific reasoning, and human compassion offer the best way to resolve these problems.

35. No one religion, church, or group should impose its views of moral conduct or legislate its version of morality for the rest of the society.

36. Human beings are capable of developing their ethical values and principles without the need of religious commandments or the benefit of clergy.

Freedom of thought:

37. Wikipedia defines freedom of thought as "the freedom of an individual to hold a view point, or thought, regardless of anyone else's view."[3] It is a fundamental human right.

38. Article 18 of the Universal Declaration of Human Rights, adopted by the General Assembly of the United Nations on December 10, 1948, proclaims the right of every human being to freedom of thought:
     Everyone has the right to freedom of thought, conscience and religion; this right includes freedom to change his religion or belief, and freedom, either alone or in community with others and in public or private, to manifest his religion or belief in teaching, practice, worship, and observance.[4]

39. Freedom of thought is the most basic of all forms of freedom.

40. Freedom of thought or expression should not be stifled by the views of the majority. Minority views must have an equal representation.

41. Because thought and language are closely linked, restricting the use of language is tantamount to restricting freedom of thought.

42. True freedom of thought breaks down prejudices and leads to enlightenment of ideas and beliefs.

Human Rights:

43. Article 1 of the United Nation's Universal Declaration of Human Rights proclaims:
     All human beings are born free and equal in dignity and rights. They are endowed with reason and conscience and should act towards one another in a spirit of brotherhood.[5]

44. Article 2 of the above declaration elaborates the above point:
     Everyone is entitled to all the rights and freedoms set forth in the Declaration, without distinction of any kind, such as race, color, sex, language, religion, political or other opinion, national or social origin, property, birth or other status. Furthermore, no distinction shall be made on the basis of the political, jurisdictional or international status of the country or territory to which a person belongs, whether it be independent, trust, non-self-governing or under any other limitation of sovereignty.

45. Any religious, philosophical, political, or ideological principle that denies any of the basic human rights, such as enumerated above, must be rejected. This should be regarded as the universal law of morality for individuals, societies, nations, and humanity at large.

Animal rights:

46. All animals possess natural rights to life, freedom, and the prohibition of torture. They should not be regarded as property by human beings.

47. Animal rights are not understood to be legal rights but rather moral obligations based on human feelings of compassion and concern for other creatures.

48. Just like torturing another person for pleasure is immoral, in the same way it is immoral to torture an animal. Such a morality is based on the fact that humans can feel the pain and suffering of others, including that of animals.

49. The morality of showing kindness to animals arises out of the same feelings that fill the hearts of humans toward other humans.

50. It is a scientifically established fact that those who are cruel to animals are also cruel to humans. Those who respect human life also respect animal life.

51. It is morally wrong to kill or torture animals for entertainment.

52. It is wrong in principle to kill animals for food when it is not essential for our survival.

53. Enlightened believers of scriptures interpret the phrase "dominion over animals" as "stewardship" rather than a license to use animals as we see fit. Whether you believe in the scriptures or not, showing kindness to animals is a humane thing to do.

Planet Earth:

54. You cannot respect life without respecting the mother that brings it forth and sustains it. The earth is the mother of all living things and therefore commands motherly respect. As Shakespeare said:

The earth that's nature's mother is her tomb. What is her burying grave, that is her womb. (*Romeo and Juliet*, act 2, scene 3).

55. Metaphorically, the earth is called Gaia, the Greek goddess of Earth. James Lovelock explains the concept of Gaia as "a complex entity involving the Earth's biosphere, atmosphere, oceans, and soil; the totality constituting a feedback or cybernetic system which seeks an optimal physical and chemical environment for life on this planet."[6]

56. An organization called the Gaia Preservation Coalition has enunciated its principles for promoting harmony within Gaia so that "all life and future generations can enjoy and share the fruits of this unique planet." It defines the metaphor of Gaia as the notion of web-like interconnectedness—everything affects everything else. It relates the emerging episteme wherein Humanity is viewed as inherently part of the Ecosphere—a self-regulating, living system, not a "resource" that can be exploited with impunity. This awareness of Humanity's embeddedness within Gaia goes under the name "Gaian consciousness." … This movement can/must lead to a new relationship between politics and governance, Humanity and the environment.[7]

57. The following words of a member of the Wintu Native American people present a sad commentary on modern man's lack of concern for the environment (substitute "modern" for "White," if you like):
The White people never cared for land or deer or bear. When we Indians kill meat, we eat it all up. When we dig roots we make little holes; when we build houses, we make little holes. When we burn grass for grasshoppers, we don't ruin things. We shake down acorns and pine nuts. We don't chop down the trees. We only use dead wood. But the White people plow up the ground, pull down the trees, kill everything. The tree says, 'Don't. I'm sore. Don't hurt me.' But they chop it down and cut it up. The Spirit of the land hates them. They blast out trees and stir it up to its depths. They saw up the trees. That hurts them. The Indians never hurt anything, but the White people destroy all. They blast rocks and scatter them on the ground. The rock says: 'Don't. You are hurting me.' But the White people pay no attention—How can the Spirit of the Earth like the White man?—Everything the White man has touched it, it is sore. (A member of the Wintu Native American people, quoted in *American Indian Environments*, p. 32).[8]

58.   Man's control over the earth is restricted, say the scriptures:
The land shall not be sold for ever: for the land is mine, for ye are strangers and sojourners with me. And in all the land of your possession ye shall grant a redemption for the land. (Lev. 25:23–24)
Eat of their fruit in their season, but render the dues that are proper on the day that the harvest is gathered. But waste not by excess for Allah does not love the wasters. (Qur'an 6:141)

59.   Human destiny is intimately linked with the survival and preservation of Mother Earth.

Human destiny:

60.   The ultimate destiny of man is a great mystery that has been the subject of religions, philosophy, and science through the ages. However, the mystery will remain unsolved as long as humans keep on believing in speculation or substituting it for a greater mystery like some religions have done.

61.   If we contemplate on human destiny (without supernatural beliefs), we can visualize the future from the experiences of the past. We know pretty clearly how human beings evolved, and history tells us how civilizations came and went. We know what helps humanity and what destroys it. Human nature is basically the same, and the same laws of the universe are forever in operation. So, it is through the knowledge gained in the past and in the present that we can throw light on the future.

62.   We know that evolution operates through its law of survival of the fittest. In the case of man, being fit means not only physically fit but also intellectually fit. In the early ages, the fitness balance tilted more on the physical side while in modern times intellect has more weight. Thus, the destiny of man will be determined largely on his intellectual prowess.

63.   We also know that physical survival is not the only goal of humanity. If man has any control over his destiny, survival with dignity, justice, freedom, happiness, and fulfillment of his potentials must be his paramount goal. This means the law of the survival of the fittest must be tempered with man's intellectual and moral powers.

64. Man has evolved physically, culturally, and morally. But there is no master plan to follow and no particular human destiny to anticipate. We can guide evolutionary forces to operate beneficially.

65. Beware of those who would like to force their ideals of human condition on others.

66. Evolutionary forces, some natural and some man-made, will no doubt determine human destiny. No one knows what our final destiny is.

67. "Dust thou art, and unto dust shalt thou return"[8] means to me that we are all formed out of atoms and molecules of the universe, and we will become atoms and molecules of the universe after death. Ultimately, our destiny is the destiny of our universe.

NOTES AND CITATIONS

Chapter 3

1. Webster's New World Dictionary, Second College Edition (New York: The World Publishing Company, 1970).

Chapter 4

1. When matter that is held together by gravity is pulled apart, it takes energy to overcome gravity. This energy is locked up in the gravitational field as a negative energy.

2. H. Genz, *Nothingness* (Reading, MA: Perseus Books, 1999).

Chapter 5

1. Leon Lederman, *The God Particle* (Boston, New York: Houghton Mifflin Company, 1993).

2. In order to conserve the total number of particles in the universe, a particle (fermion) cannot be created without also creating its antiparticle.

3. P. W. Milonni, R. J. Cook, and M. E. Goggin, Radiation pressure from the vacuum: physical interpretation of the Casimir force, *Physical Review.* (1988):A 38, 1621.

4. W. E. Lamb and R. C. Retherford, Fine structure of the hydrogen atom by a microwave method, *Phys. Rev.* (1947):72, 241.

5. E. P. Tyron, Is the universe a vacuum fluctuation? *Nature* (1973):246, 396.

6. Professor Stephen Hawking Lectures,
   http://www.admin.ox.ac.uk/po/news/2005-06/feb/27.shtml.

7. Wikipedia, the free encyclopedia, "Alan Guth,"
   http://en.wikipedia.org/wiki/Alan_Guth.

Chapter 6

1. Lederman, *The God Particle*.

2. A blackbody is defined in physics as an object that absorbs all radiation incident upon it. For that reason, it is a perfect emitter of heat radiation whose wavelength is correlated with its temperature.

3. George Gamow, *My Worldline* (New York: Viking Adult, 1970).

Chapter 7

1. J. D. Barrow, *The Origin of the Universe* (New York: Basic Books, Harper Collins Publishers, 1994).

2. In 2006, the International Astronomical Union classified Pluto as a dwarf planet.

3. The CMBR data is from the universe when it was about four hundred thousand years old. This is the time when the universe first became transparent to radiation following the big bang.

4. The relationship involves Hubble's constant H for a given cosmological time and the gravitational constant G. The critical density is given by $3H^2/(8\pi G)$. Taking the best available values of $H = 75$ Km s^{-1}/Mpc $= 2.43 \times 10^{-18}$ s^{-1} and $G = 6.67 \times 10^{-11}$ Nm^{2}kg^{-2}, we get critical density $= 1.06 \times 10^{-26}$kg/m^3.

5. T. S. Eliot, "The Hollow Men," http://www.cs.umbc.edu/~evans/hollow.html.

Chapter 8

1. Amino acids (e.g., glycine, leucine, and methionine) are basic molecules that can join together in sequences to make proteins. The sequence to create a particular protein is dictated by nucleotide base sequences in the DNA genetic code.

2. P. Whitfield, *From So Simple a Beginning* (New York: MacMillan Publishing Company, 1993), 18.

3. Ibid., 84.

4. Stephen Hawking, *A Brief History of Time* (London: Bantam Press, 1988).

5. Paul Davies, *Superforce: The Search for a Grand Unified Theory of Nature* (New York: Simon and Schuster Inc., 1985).

6. Brandon Carter, 1974. Large number coincidences and the anthropic principle, *IAU Symposium 63: Confrontation of Cosmological Theory with Astronomical Data* (1974), http://en.wikipedia.org/wiki/Brandon_Carter.

7. J. D. Barrow and F. J. Tipler, *The Anthropic Cosmological Principle* (Oxford: Oxford University Press, 1986).

8. P. E. Johnson, *Darwin on Trial* (Washington DC: Regnery Gateway, 1991), 00.

9. M. J. Behe, *Darwin's Black Box: The Biochemical Challenge to Evolution* (New York: Free Press, 1996).

10. W. A. Dembski, *Intelligent Design: The Bridge Between Science and Theology* (Nottingham: InterVarsity Press, 1999).

11. P. Whitfield, *From So Simple a Beginning* (New York: MacMillan Publishing Company, 1993), 91.

Chapter 9

1. P. Boyer, Religious thought and behavior as by-products of brain function, *Trends in Cognitive Sciences* 7 (2003):119–24.

2. Karl Marx, *Critique of Hegel's Philosophy of Right (1843)* (Cambridge: Cambridge University Press, 1977).

3. R. Stark and W. S. Bainbridge, *A Theory of Religion* (New Jersey: Rutgers University Press, 1996).

4. R. T. Firth, *Religion* (New York: Routledge, 1995).

5. Ibid.

6. W. W. Howell, *The Heathens* (Garden City, NY: Doubleday, 1948).

Chapter 10

1. BCE stands for before the Common Era (see note 1 of chapter 11).

2. Ernest Valaea, The ultimate reality in world religions, 1999–2007, http://www.comparativereligion.com/god.html.

3. Trevor Barnes, *Religions* (New York: Kingfisher Publications, 1999), 61.

4. Ibid., 50.

Chapter 11

1. The date of birth of Jesus Christ is uncertain. The Christian era begins with the year formerly thought to be that of the birth of Jesus, but it is now thought that he was probably born between 8–4 BC. The terms BC (before Christ) and AD (*anno Domini* or in the year of the Lord) mark dates before and after the birth of Christ. The Common Era is the Christian era that corresponds to the current dates. Some writers prefer the use of the term BCE. (before Common Era) instead of BC to avoid confusion over the birth date of Christ. Similarly, CE is used instead of AD.

Chapter 12

1. Moojan Momen, *The Phenomenon of Religion* (Oxford: One World, Oxford Press, 1999), 439.

2. Ibid., 437.

Chapter 13

1. Wikipedia, the free encyclopedia, "Agnosticism," http://en.wikipedia.org/wiki/agnosticism.

2. Ibid.

3. Famous atheists, freethinkers, deists, and agnostics, http://www.wonderfulatheistsofcfl.org/Quotes.htm.

4. Ibid.

5. Ibid.

6.   Ibid.

7.   Ibid.

8.   American Humanist Association, *Manifesto I, II, III* (1933, 1973, 2003), http://www.americanhumanist.org.

9.   Council for Secular Humanism, *A Secular Humanist Declaration* (1980), http://www.secularhumanism.org/index.php?section=mainpage=declaration.

10. American Humanist Association, *Humanist Manifesto I (1993)*, http://www.americanhumanist.org/about/manifesto1.html.

Chapter 14

1.   Reported by CBC.ca from the journal *Nature* (April 6, 2006), http://www.cbc.ca/story/science/national/2006/04/05/fossil-fish-20060405.html?reff=rss.

2.   Reported by CNN.com, http://www.cnn.com/2006/TECH/science/04/12/fossil.evolution.ap/index.html.

Chapter 15

1.   Religious Tolerance.org, http://www.religioustolerance.org/rel plur.htm.

2.   Ibid.

3.   Wikipedia, the free encyclopedia, "Enlightenment," http://en.wikipedia.org/wiki/Enlightenment(concept).

4.   Blaise Pascal, *Pascal's Pensees*, Trans. W. F. Trotter (1910), http://plato.stanford.edu/entries/pascal-wager/#Bib.

5.   World scripture Web site, "Hell," http://www.unification.net/ws/theme044.htm.

6.   Momen, *The Phenomenon of Religion*, 234.

7.   Eugene Khutoryansky, Objective morality based on scientific and rational reasoning, 1996, http://members.aol.com/okhutor/essay/morals.html.

Chapter16

1. Famous atheists, freethinkers, deists, and agnostics, http://www.wonderfulatheistsofcl.org/Quotes.htm.

2. Whitfield, *From So Simple a Beginning*, 91.

3. Wikipedia, the free encyclopedia, "Freedom of thought," http://en.wikipedia.org/wiki/Freedom_of_thought.

4. Web site devoted to the Fifth Anniversary of the Universal Declaration of Human Rights (1948–1998), http://www.un.org/Overview/rights.html.

5. Ibid.

6. Wikipedia, the free encyclopedia, "Gaia hypothesis," http://en.wikipedia.org/wiki/Gaia_hypothesis.

7. Gaia Preservation Coalition, "Gaia: the metaphor" and "Gaia: the movement," http://www.gaiapc.ca/.

8. Momen, *The Phenomenon of Religion*, 357.

9. Gen. 3:19 (King James version).

BIBLIOGRAPHY

American Humanist Association. "Humanist Manifesto I (1933)." http://www.americanhumanist.org/humanism/manifesto1.php (accessed March 17, 2006).

American Humanist Association, "Humanist Manifesto II (1973)." http://www.americanhumanist.org/about/manifesto2.php (accessed March 17, 2006).

Barnes, Trevor. *Religions*. New York: Kingfisher, 1999.

Barrow, John D. *The Origin of the Universe*. New York: BasicBooks/Harper Collins Publishers, Inc., 1994.

The Holy Bible (King James version). Chicago: The John A. Hertel Co., 1963.

Boyer, Pascal. 2003. Religious thought and behavior as by-products of brain function. *Trends in Cognitive Science* 7:119–124.

Calle, Carlos I. *Superstrings and Other Things*. Philadelphia: Institute of Physics Publishing, 2001.

Charap, John M. *Explaining the Universe*. New Jersey: Princeton University Press, 2002.

Council for Secular Humanism. "A Secular Humanist Declaration (1980)." http://www.secularhumanism.org/index. php?section=main&page=declaration (accessed March 18, 2006).

Davies, Paul. *The Mind of God*. New York: Simon Schuster, 1992.

Davies, Paul. "What Happened Before the Big Bang?" http://www.fortunecity.com/emachines/e11/86/big-bang.html (accessed September 23, 2004).

Dunkelberg, Pete. "Irreducible Complexity Demystified." http://www.talkdesign.org/faqs/icdmyst/ICDmyst.html (accessed April 4, 2006).

Einstein, Albert. *The World As I See It*. New York: Philosophical Library, Inc., 1949.

Einstein, Albert. *Relativity: The Special and the General Theory*. New York: Crown, 1961.

Ferris, Timothy. *The Whole Shebang*. New York: Simon & Schuster, 1997.

Filkin, David. *Stephen Hawking's Universe*. New York: BasicBooks/Harper Collins Publishers, Inc., 1997.

Ford, Lawrence H., and Thomas A. Roman. "Negative Energy, Wormholes and Warp Drive." http://www.physics.hku.hk/~tboyce/sf/topics/wormhole/wormhole.html (accessed October 17, 2004).

French Animal Rights League, "Universal Declaration of Animal Rights." http://league-animal-rights.org/en-duda.html (accessed June 21, 2006).

Genz, Henning. *Nothingness*. Reading, MA: Perseus Books, 1998.

Green, Alex. "The Science and Philosophy of Consciousness." http://www.users.globalnet.co.uk/~lka/conr.htm.

Greene, Brian. *The Fabric of the Cosmos*. New York: Alfred A. Knopf, 2004.

Harris, Sam. *The End of Faith*. New York: W. W. Norton & Company, 2004.

Hawking, Stephen. *Black Holes and Baby Universes and Other Essays*. New York: Bantam Books, 1993.

Hawking, Stephen. *A Brief History of Time*. New York: Bantam Press, 1998.

Khutoryansky, Eugene. "Objective Morality Based on Scientific and Rational Reasoning." http://members.aol.com/okhutor/essay/morals.html (accessed May 27, 2006).

Lederman, Leon. *The God Particle*. New York: Houghton Mifflin Company, 1993.

Levy, David H., ed. *Cosmos*. New York: St. Martin's Press.

Momen, Moojan. *The Phenomenon of Religion*. Oxford: Oneworld Publications, 1999.

Morris, Richard. *Cosmic Questions*. New York: John Wiley& Sons, Inc., 1993.

Narlikar, Jayant. *The Lighter Side of Gravity*. Cambridge: Cambridge University Press, 1996.

Nussbaum, Martha C. "Morality and Emotions." http://www.geocities. com/Athens/Rhodes/3724/Cytrix/cdrom5/Routledge morality emotion.htm?200 (accessed May 28, 2006).

Parsons, Paul. *The Big Bang*. BBC Worldwide Publishing Ltd/DK Publishing, Inc., 2001.

Physics Web, "The Casimir Effect: a Force from Nothing." http:// physicsweb.org/articles/world/15/9/6 (accessed October 17, 2004).

The Qur'an. Translated by Abdullah Yusuf Ali. New York: Tahrike Tarsile Qur'an, Inc., 1999.

Rees, Martin. *Our Cosmic Habitat*. Princeton: Princeton University Press, 2001.

Reeves, Hubert, Joel De Rosnay, Yves Coppens, and Dominique Simonnet. *Origins: Cosmos, Earth, And Mankind*. New York: Arcade Publishing, 1996.

Rowe, Stan. "The Living Earth and its Ethical Priority." http://www. ecospherics.net/pages/Roweliving.htm (accessed June 22, 2006).

Sagan, Carl. *Cosmos*. New York: Random House, Inc., 1980.

Spence, Pam, ed. *The Universe Revealed*. Cambridge: Cambridge University Press, 1999.

United Nations. "Universal Declaration of Human Rights (1948)." http://www.un.org/Overview/rights.html (accessed June 19,2006).

Valea, Ernest. "The Ultimate Reality in World Religions." http://www. comparatiereligions.com/god.html (accessed February 21, 2005).

Watts, Charles. "The Origin, Nature, and Destiny of Man." http://infidels.org/library/historical/charles watts/origin.html (accessed June 26, 2006).

Whitfield, Philip. *From So Simple a Beginning.* New York: Macmillan Company, 1993.

Zukav, Gary. *The Dancing Wu Li Masters.* New York: William Morrow and Company, Inc., 1979.

INDEX

978-0-595-43006-2
0-595-43006-6

9 780595 430062